Patrick Huff

CRESST-II Dark Matter Experiment

Patrick Huff

CRESST-II Dark Matter Experiment

The Detector Parameters Determining the Sensitivity of the CRESST-II Experiment

Südwestdeutscher Verlag für Hochschulschriften

Impressum / Imprint

Bibliografische Information der Deutschen Nationalbibliothek: Die Deutsche Nationalbibliothek verzeichnet diese Publikation in der Deutschen Nationalbibliografie; detaillierte bibliografische Daten sind im Internet über http://dnb.d-nb.de abrufbar.

Alle in diesem Buch genannten Marken und Produktnamen unterliegen warenzeichen-, marken- oder patentrechtlichem Schutz bzw. sind Warenzeichen oder eingetragene Warenzeichen der jeweiligen Inhaber. Die Wiedergabe von Marken, Produktnamen, Gebrauchsnamen, Handelsnamen, Warenbezeichnungen u.s.w. in diesem Werk berechtigt auch ohne besondere Kennzeichnung nicht zu der Annahme, dass solche Namen im Sinne der Warenzeichen- und Markenschutzgesetzgebung als frei zu betrachten wären und daher von jedermann benutzt werden dürften.

Bibliographic information published by the Deutsche Nationalbibliothek: The Deutsche Nationalbibliothek lists this publication in the Deutsche Nationalbibliografie; detailed bibliographic data are available in the Internet at http://dnb.d-nb.de.

Any brand names and product names mentioned in this book are subject to trademark, brand or patent protection and are trademarks or registered trademarks of their respective holders. The use of brand names, product names, common names, trade names, product descriptions etc. even without a particular marking in this work is in no way to be construed to mean that such names may be regarded as unrestricted in respect of trademark and brand protection legislation and could thus be used by anyone.

Verlag / Publisher:
Südwestdeutscher Verlag für Hochschulschriften
ist ein Imprint der / is a trademark of
OmniScriptum GmbH & Co. KG
Heinrich-Böcking-Str. 6-8, 66121 Saarbrücken, Deutschland / Germany
Email: info@svh-verlag.de

Herstellung: siehe letzte Seite /
Printed at: see last page
ISBN: 978-3-8381-1620-4

Zugl. / Approved by: München, TU, Diss., 2010

Copyright © 2010 OmniScriptum GmbH & Co. KG
Alle Rechte vorbehalten. / All rights reserved. Saarbrücken 2010

Max-Planck-Institut für Physik
Werner-Heisenberg-Institut

The Detector Parameters Determining the Sensitivity of the CRESST-II Experiment

Patrick Huff

March 2010

Dissertation
an der Fakultät für Physik
der Technischen Universität München

TECHNISCHE UNIVERSITÄT MÜNCHEN

Max-Planck-Institut für Physik
Werner-Heisenberg-Institut

The Detector Parameters Determining the Sensitivity of the CRESST-II Experiment

Patrick Huff

Vollständiger Abdruck der von der Fakultät für Physik der Technischen Universität München zur Erlangung des akademischen Grades eines Doktors der Naturwissenschaften genehmigten Dissertation.

Vorsitzender: Univ.-Prof. Dr. A. Ibarra
Prüfer der Dissertation: 1. Hon.-Prof. A. C. Caldwell, Ph.D.
                            2. Univ.-Prof. Dr. T. Lachenmaier

Die Dissertation wurde am 08.03.2010 bei der Technischen Universität München eingereicht und durch die Fakultät für Physik am 23.04.2010 angenommen.

Zusammenfassung

Die Funktionsweise der Detektormodule des CRESST-II Experiments wird im Rahmen dieser Arbeit diskutiert und das Verständnis dieser weiter entwickelt. CRESST-II ist ein Experiment, dass direkt nach Dunkler Materie in Form von WIMPs sucht.
Indirekte Beobachtungen Dunkler Materie wie z.B. Rotationskurven von Galaxien, welche die grundlegende Motivation für das gesamte Experiment bilden, werden in Abschnitt 1.1 präsentiert. Diese bis heute unverstandenen Beobachtungen können durch unterschiedliche Erweiterungen des Standardmodells der Teilchenphysik erklärt werden (Abschnitt 1.2). Die von einigen dieser Theorien vorhergesagten WIMPs sind eine der bevorzugten Möglichkeiten, die Natur der Dunklen Materie zu erklären. Möglichkeiten für einen direkten Nachweis dieser Teilchen werden in Abschnitt 1.3 diskutiert.
CRESST-II, welches eines der Experimente ist, das direkt nach WIMPs sucht, wird in Kapitel 2 präsentiert. Ausgehend von der CRESST-II-Funktionsweise, welche in Abschnitt 2.2 beschrieben wird, können die wesentlichen Parameter verstanden und bestimmt werden, welche die Sensitivität von CRESST-II festlegen (Abschnitt 2.3). Dies sind die sogenannten Quenchingfaktoren und die Energieauflösungen der beiden gemessen Signale.
Die Quenchingfaktoren werden in Kapitel 3 behandelt. Zunächst werden in Abschnitt 3.1 aktuelle Daten präsentiert. Im darauf folgenden Abschnitt 3.2 werden diese Werte anhand einer mikroskopischen Erklärung untereinander in Verbindung gebracht. Schlüsse und Folgerungen aus dieser Erklärung sind abschließend in Abschnitt 3.3 dargelegt.
In Kapitel 4 werden die Energieauflösungen der beiden gemessenen Kanäle diskutiert. Die Parameter, welche die gemessene Signalhöhen bestimmen, werden in Abschnitt 4.2 aufgelistet. Abschnitt 4.3 diskutiert hingegen die Parameter, welche im wesentlichen für die Messunsicherheiten verantwortlich sind.
Um den Zusammenhang der physikalischen Abläufen und den physikalischen Eigenschaften eines Lichtdetektors verstehen zu können, welche wesentlich die Sensitivität von CRESST-II beeinflussen, wird in Kapitel 5 ein Modell eingeführt und angepasst. Die mathematische Lösung wird in Abschnitt 5.4 präsentiert. Sie wird danach in Abschnitt 5.5 diskutiert und physikalisch interpretiert.
Abschließend befassen sich Kapitel 6 und 7 mit den zuvor bestimmten Parametern, welche die Sensitivität des Experiments bestimmen. Einige dieser Parameter werden modifiziert, bei anderen werden mögliche Änderungen vorgeschlagen.

Abstract

Within the framework of this thesis the functional principles of detector modules of the CRESST-II experiment are discussed and the understanding of their behavior is developed in further details. CRESST-II is an experiment searching directly for Dark Matter in the form of WIMPs.

Indirect observations of Dark Matter, e.g. galactic rotation curves, which are the motivation for the experiment, are presented in section 1.1. These up to now unexplained observations can be understood by theories extending the standard model of particle physics (section 1.2). Out of these theories, WIMPs are one of the most favored particles explaining the observations. A direct detection of these particles is discussed in section 1.3.

CRESST-II, as one experiment searching directly for WIMPs, is presented in chapter 2. From the CRESST-II operational principles described in section 2.2, one can understand the main parameters determining the experiment's sensitivity (section 2.3). These are the so-called quenching factors and the energy resolutions of the two measured channels.

The quenching factors are discussed in chapter 3. In this chapter, first, recent measured data are presented in section 3.1. Afterwards in section 3.2, these data are explained on a microscopic scale. Conclusions and outcome of this explanation are discussed in section 3.3.

In chapter 4, the energy resolutions of the two measured channels are discussed. The parameters which determine the measured signal height are itemized in section 4.2. In section 4.3, the parameters dominating the measurement uncertainty are discussed.

To be able to understand the connection between the physics in the light detector and its material properties a model is taken and adapted in chapter 5. The mathematical solution of the model is presented in section 5.4, and discussed and physically interpreted in section 5.5.

Finally, in chapter 6 and 7, some of the previously found out parameters which determine the CRESST-II sensitivity are modified and further possible variations are discussed.

Contents

1 Dark Matter   **1**
1.1 Indirect Observations of Dark Matter 1
1.2 Theoretical Explanations 6
    1.2.1 Dark Matter Properties 6
    1.2.2 Dark Matter Candidates 6
1.3 Direct Detection of Dark Matter in the Formof WIMPs 8
    1.3.1 Interaction Rate 9
    1.3.2 Deposited Energy 11
    1.3.3 Energy Spectrum 11
    1.3.4 WIMP-Signature 14

2 The CRESST-II Experiment   **17**
2.1 CRESST-II Setup 17
    2.1.1 Background Reduction and Identification 17
    2.1.2 CRESST-II Detector Modules 22
2.2 CRESST-II Operation 24
    2.2.1 Signal Measurement 24
    2.2.2 Signal Differentiation 26
2.3 CRESST-II Sensitivity 29
    2.3.1 Light-Phonon-Plane 29
    2.3.2 Parameters of the CRESST-II Sensitivity 34
2.4 Motivation of this Work 36

3 Quenching Factors of $CaWO_4$ **37**
3.1 Quenching Factors 37
    3.1.1 Quenching Factor Definition 37
    3.1.2 Quenching Factor Measurement for $CaWO_4$ 38
3.2 Quenching Factor Explanation 41
    3.2.1 Saturation of the Light Production 41
    3.2.2 Linearity of the Light Production 43
    3.2.3 Energy Deposition and Light Production 45
3.3 Energy Dependence of the Bands 48

4 Energy Resolution of the Light Channel — 51

- 4.1 Setup of the Light Channel — 53
  - 4.1.1 Target Crystal — 53
  - 4.1.2 Reflecting Housing — 55
  - 4.1.3 Light Absorber — 55
  - 4.1.4 Thermometer — 55
- 4.2 Mean Measured Energy — 56
  - 4.2.1 Energy Transformation: p — 58
  - 4.2.2 Photon Transport: q — 59
  - 4.2.3 Phonon Transport: r — 61
  - 4.2.4 Heat Capacity: C_T — 63
  - 4.2.5 Transition Slope: m — 64
  - 4.2.6 Thermometer Bias Current: I_T — 66
  - 4.2.7 Shunt Resistance: R_S — 67
  - 4.2.8 Thermometer Resistance: R_T — 67
  - 4.2.9 Current to Voltage Transformation: $\frac{\partial U}{\partial \Phi_0} \frac{\partial \Phi_0}{\partial I}$ — 67
  - 4.2.10 Energy Calibration: c — 68
- 4.3 Measurement Uncertainty — 69
  - 4.3.1 Energy Transformation: δp — 69
  - 4.3.2 Photon Transport: δq — 69
  - 4.3.3 Absorbed Energy: δE_{LD} — 70
  - 4.3.4 Phonon Transport: δr — 70
  - 4.3.5 Heat Capacity: δC_T — 70
  - 4.3.6 Thermometer Temperature Rise: $\delta\left(\Delta T_T^{dep}\right)$ — 71
  - 4.3.7 Transition Slope: δm — 71
  - 4.3.8 Thermometer Resistance Change: $\delta(\Delta R_T)$ — 71
  - 4.3.9 SQUID Bias Current Change: $\delta(\Delta I_S)$ — 73
  - 4.3.10 Summing Up and Conclusions — 74

5 Model of the Energy Transport in the Light Detector — 77

- 5.1 Introduction — 77
- 5.2 Model Assumptions — 78
- 5.3 Equations for the Temperatures — 81
  - 5.3.1 Basic Equations for the Temperatures — 81
  - 5.3.2 Thermometer Heater — 83
  - 5.3.3 Bias Current Self-Heating — 86
  - 5.3.4 Phonon Collectors — 88
  - 5.3.5 Time Duration of the Energy Deposition — 90
  - 5.3.6 Equations for the Relative Temperatures — 93
- 5.4 Solution of the Equations — 93
- 5.5 Discussion of the Solution — 95
  - 5.5.1 First Case: $G_{AT} \equiv 0$ — 96
  - 5.5.2 Second Case: $G_{AT} \gg G_{BT}^* = G_{AB}$ — 97
  - 5.5.3 Third Case: $C_T \ll C_A$ & $G_{AB} \equiv 0$ — 99
  - 5.5.4 Summary — 101
  - 5.5.5 Fourth Case: Real Light Detector — 102

6	**Light Detector Optimization**		**111**
	6.1 Experiments		111
		6.1.1 Black Silicon Light Absorber	111
		6.1.2 Separated Heater	115
	6.2 Discussion of Modifications		120
		6.2.1 Thermometer Size	120
		6.2.2 Maximization of ΔT_e	122
7	**Conclusions and Perspectives**		**125**
A	**Heat Capacities**		**129**
B	**Solution of the Two Coupled Differential Equations**		**133**
C	**Practical Remarks**		**137**
	C.1 Superconducting Cooling Strip		137
	C.2 Destroyed Gold Structure		139
	C.3 Thermometer Temperature Increase During a Pulse		140
D	**Notation**		**143**
	Bibliography		**146**

Chapter 1

Dark Matter

Although the nature of more than 95 % of the matter and energy in the universe are unknown, a rough idea of the constitution of our universe is well-established. It consists mainly of Dark Energy (72.6 %) and Dark Matter (22.8 %). Only about 4.6 % of the total universe density are well-known baryonic matter [1]. The nature of Dark Energy and Dark Matter is unknown.

For Dark Energy up to now no conclusive idea exists, which could explain its nature. For **Dark Matter**, on the other hand, different concrete ideas have been developed, which could explain it (section 1.2). In the next section, some of the most convincing indirect observations of Dark Matter will be presented shortly. These indirect observations give the motivation for the direct detection of Dark Matter (section 1.3) which many ongoing experiments are aiming for.

1.1 Indirect Observations of Dark Matter

Up to now Dark Matter could only be observed indirectly. Some of these observations are presented in the following. Although they use different methods and search on different cosmological scales all these observations have the same conclusion:

> There is much more Dark Matter than luminous matter in the universe.

Rotation Curves of Spiral Galaxies

The stars of spiral galaxies rotate around their galactic center. If a galaxy consisted mainly of the visible stars, the rotation speed on stable Keplerian orbits should be $v(r) \sim 1/\sqrt{r}$ for matter at radial distances beyond the visible stars. Instead, the rotation speed $v(r)$ is approximately constant for huge radii r. I.e. the outer matter moves faster than the gravitational binding of the visible objects would allow for.

In figure 1.1 such a measured rotation curve of the spiral galaxy M33 is shown (points). The velocity contribution from the stellar disk (short dashed line) together with the gas contribution (long dashed line) cannot explain the measured

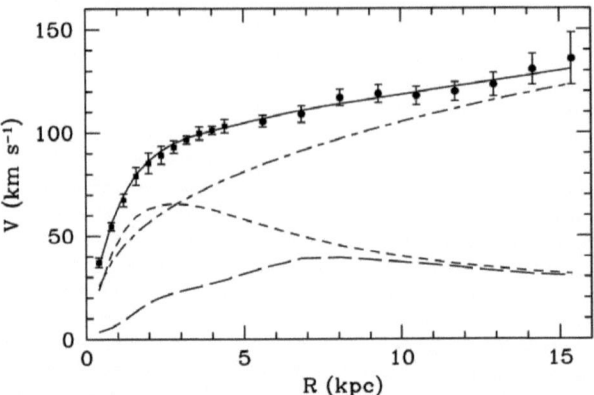

Figure 1.1: The radial velocity distribution of spiral galaxy M33 is shown. Measured values of the rotation curve are plotted with points. The velocity contribution from the stellar disk of M33 is shown with the short dashed line, the gas contribution with the long dashed line. Without an additional halo contribution (dashed-dotted line) the measured velocities cannot be explained. Figure taken from [2].

velocities. Only with an additional contribution from a halo (dashed-dotted line), which contains large amounts of Dark Matter, the measurement can be explained [2].

Velocity Dispersions of Dwarf Galaxies and Galaxy Clusters

In galaxy clusters and especially in dwarf galaxies huge amounts of Dark Matter are expected. The reason for this is that the observed matter velocities are unusually large, which are not allowed from the gravitational binding of the luminous matter only [3].

In a gravitationally bound system which is in equilibrium, the virial theorem expects:

$$2 \langle E_{Kin} \rangle = -\langle E_{Pot} \rangle$$

The averaged kinetic energy can be determined by the total mass M, the total luminosity L, the individual luminosities l_i, and the individual peculiar velocities v_i of the constituent parts of the system:

$$\langle E_{Kin} \rangle = \langle E_{Kin} \rangle (M, L, l_i, v_i)$$

On the other hand, the averaged potential energy depends on the total mass M, the total radius R, and the mass distribution which is described by α:

$$\langle E_{Pot} \rangle = \langle E_{Pot} \rangle (M, R, \alpha)$$

1.1 Indirect Observations of Dark Matter

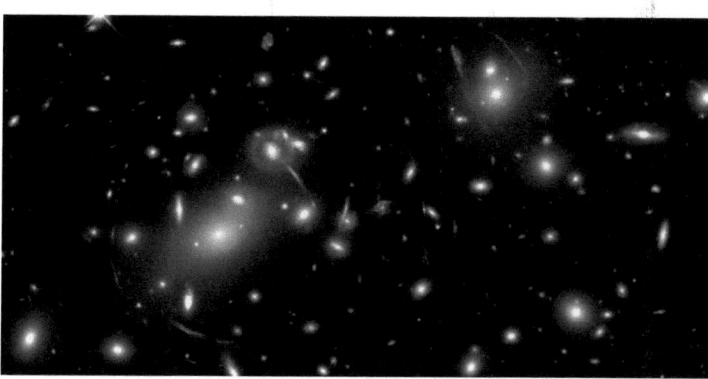

Figure 1.2: A picture of the galaxy cluster Abell 2218 from Hubble Space Telescope. The cluster acts like a lens which distorts background galaxies into long thin arcs around this galaxy cluster. The picture is taken from [5].

Comparing these energies a significant discrepancy can be seen [4]. The observations can be explained by the presence of a Dark Matter component in the systems largely exceeding the luminous matter content.

Gravitational Lensing

The mass distribution, of for example a galaxy cluster, works like a gravitational lens bending the path of the light. If the mass is perfectly symmetric (e.g. a black hole) and exactly on the line of sight, the light of an object behind the gravitational lens can be seen as a ring. Such a ring is called Einstein ring. In reality, it happens much more often that only a part of such a ring can be observed. These parts are called arcs.

An example for arcs can be seen in figure 1.2, where a picture of galaxy cluster Abell 2218 is shown. The long arcs are centered around Abell 2218. They are the pictures of mainly the same luminous matter from far behind the galaxy cluster. From these arcs, the mass distribution of the lensing galaxy cluster can be derived. Again a huge amount of Dark Matter is necessary in the lensing system to explain the observations, e.g. [6] or [7]. The mass distributions inferred from gravitational lensing are generally in good agreement with the expected Dark Matter amount deduced from other observations.

Temperatures of Gases in Galaxy Clusters

Observations of galaxy clusters show that they contain huge amounts of hot gas that exceed the luminous mass by many times. But still this mass is not enough to explain the observation that, despite its temperature and therefore its kinetic energy, the hot gas is bound in the system. Huge amounts of Dark Matter can explain this phenomenon. By analyzing the gas profile even a profile of the

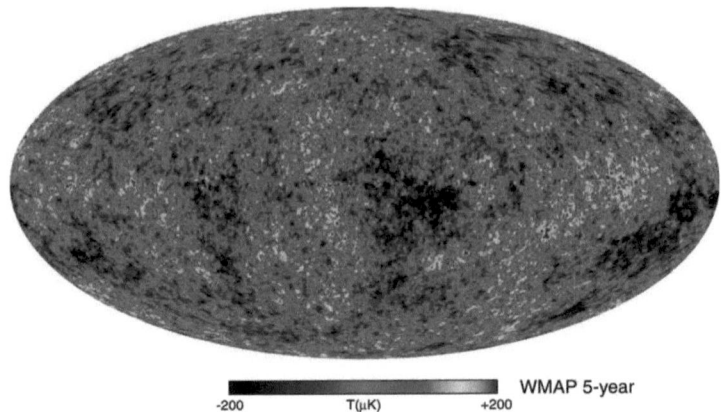

Figure 1.3: This figure shows the temperature fluctuations of the CMB. It provides information about the constitution of the universe. Figure taken from [9].

expected Dark Matter distribution can be derived [8].

Temperature Fluctuations of the CMB

The Cosmic Microwave Background radiation (CMB) is the omnipresent thermal radiation in the universe. Its temperature today is 2.725 K. The measurement of the CMB is not only an indication for the Big Bang theory,[1] it provides also an information about the constitution of the universe.
About 400 000 years after the Big Bang photons decoupled from the cosmic plasma, which combined to neutral atoms, so from that time on the photons could travel through the universe nearly without interactions. These photons, the CMB, have not been interacting since they decoupled from the ordinary matter. Small temperature fluctuations of the CMB therefore reflect the density fluctuations 13.7 billion years ago, see figure 1.3. The analysis of these fluctuations let expect that about 21.4 % of the universe mass are in form of the Dark Matter [9].

Modeling the Universe Structures

Another observation which indicates the existence of Dark Matter, is the overall distribution of the visible matter in our universe. The observed matter distribution can be compared with results of simulations. If they do not fit, then the assumption of the simulation cannot be right; if they fit, then the assumptions are maybe correct.
In general, simulations of the evolution of the universe begin about 400 000 years

---

[1]The main two other observations suggesting the Big Bang theory are Hubble's law, i.e. the expansion of space and the primordial nucleosynthesis, i.e. the mechanism which explains today's abundance of the light elements in the universe.

1.1 Indirect Observations of Dark Matter

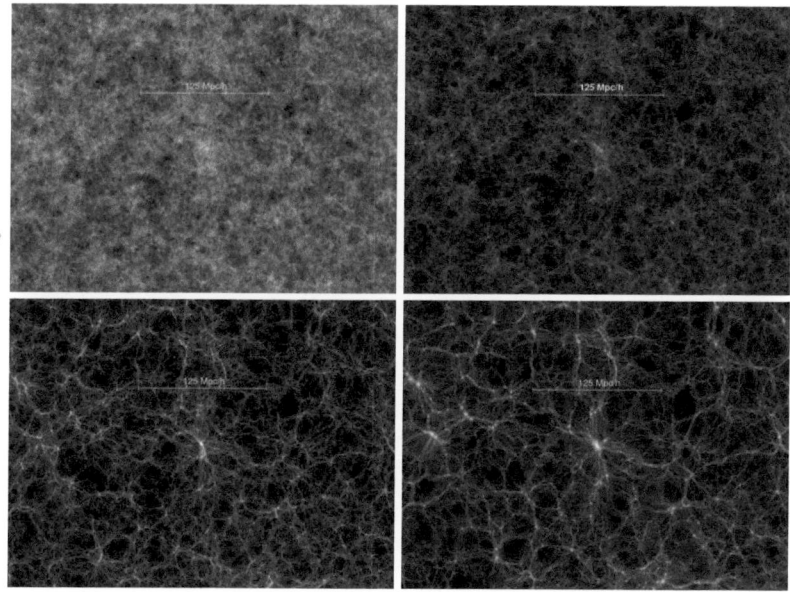

Figure 1.4: Four different points in time of the Millennium Simulation [10]. The first picture on the top left is at $t = 0.21\,\text{Gyr}$ after the Big Bang. The second one on the top right is at $t = 1.0\,\text{Gyr}$. The one on the bottom left is at $t = 4.7\,\text{Gyr}$ and the last picture on the bottom right is today at $t = 13.6\,\text{Gyr}$.

after the Big Bang. At that time the universe got transparent for photons. Only small fluctuations in the density profile are assumed. Simulations starting at that time without Dark Matter cannot reproduce the matter distribution observed today; but with Dark Matter included the simulations fit well. The reason for this is that Dark Matter clumps together earlier than baryonic matter, since it is not kept apart from each other by the electromagnetic radiation. Only with this earlier formation of the Dark Matter the universe can develop to the observed matter distribution of today.

One simulation supporting this is the so-called Millennium Simulation. Four different points in time of this simulation can be seen in figure 1.4. The simulation starts 300 000 years after the Big Bang with a nearly homogeneous Dark Matter distribution. It ends with a matter distribution similar to the one observed today [11].

The simulations of the evolution of the universe show that with the existence of a significant amount of Dark Matter the observed structures of the universe can be explained.

1.2 Theoretical Explanations

As seen above, observations indicate a huge amount of Dark Matter in our universe. These independent measurements suggest that about 23 % of the universe consists of Dark Matter. That is five times more than the known matter and more than 50 times more than the luminous matter.
In the following different theoretical explanations will be introduced. In section 1.2.1 **properties** which Dark Matter candidates have to fulfill to be consistent with the observations will be specified. Afterwards, in section 1.2.2, different motivated **candidates** which have these properties will be presented.

1.2.1 Dark Matter Properties

Assume that Dark Matter consists mainly of one component. Then this component has to fulfill different properties, which follow from the indirect Dark Matter observations. Consequently, a well motivated candidate for Dark Matter should be able to fulfill the properties which are itemized in the following:

- It should be **electrically neutral**, otherwise it could emit or absorb light and would not be dark. It should be **color neutral**, not to interact strongly, and moreover it should **interact gravitationally**. On the contrary, for the weak interaction there is no constraint. A participation in the weak interaction is possible but not necessary. If Dark Matter interacts weakly, this can open the possibility for a direct detection.

- It should provide the correct relic abundance. Firstly, this means that it has to be **stable** at least on a cosmological time scale, since it existed at the beginning of the universe (structure formation, CMB) and is still observable (rotation curves, gravitational lensing). Secondly, assuming a spherical isothermal Dark Matter halo enclosing our galaxy, it should provide a **local Dark Matter density** of $0.2 - 0.4\,\text{GeV}/\text{cm}^3$ today [12].

- It should be non-relativistic at the time of structure formation, i.e. **Cold Dark Matter**:[2] Cold Dark Matter builds up small structures, whereas Hot Dark Matter builds up large structures. Simulations show that the structure formation can only yield the observed mass distribution of today when small structures form first. This is called bottom-up structure formation. Therefor a huge amount of Cold Dark Matter is necessary.

- It should finally **not** be **excluded** by any experiment.

1.2.2 Dark Matter Candidates

In this section some of the most developed and motivated Dark Matter candidates will be presented. They all fulfill the properties listed above. Whether one of these candidates is the right one could not be clarified up to now.

---

[2]Relativistic dark matter at that time is called Hot Dark Matter.

1.2 Theoretical Explanations

Out of the known particles, which are united in the standard model of particle physics, the **neutrino** is the only possible candidate for Dark Matter, since it is electrically and color neutral. The neutrino also takes part in the gravitational interaction and it is stable. However, from the known upper neutrino mass and density limits one can derive that their total mass is much too small to provide a significant contribution to Dark Matter [13]:

$$0.5\,\% < \Omega_\nu < 1.5\,\%$$

Also the theory of structure formation contradicts neutrinos being a significant amount of Dark Matter. Since neutrinos were relativistic at the time of structure formation they contribute to Hot Dark Matter. This supports the top-down structure formation, which cannot be supported by observations.
Out of this can be concluded:[3]

> **The existence of Dark Matter is evidence for new physics beyond the standard model of particle physics.**

The standard model particles cannot explain the Dark Matter observations. However, there exist many different theories, which extend the standard model. And still many of these theories exist, which let expect new, stable, but up to now unobserved particles. Some of these particles are proper candidates for building up Dark Matter. For example the **Little Higgs Model** [14] or the **Kaluza–Klein theory** [15] can predict such particles. Also **primordial black holes** [16] or very **light axions** (μeV - meV) [17] could be the main contribution of Dark Matter. Theories like these predict particles, which could explain the nature of the indirect observed Dark Matter.

However, the most developed extension, which provides such a particle, is **Supersymmetry**. One of the properties of this theory is, that each particle of the standard model has a superpartner. The superpartner of a boson is a fermion and vice-versa. Additionally in this theory each particle of the standard model has an R-parity of +1 and each supersymmetric particle has an R-parity of -1. If the R-parity in Supersymmetry is conserved, then the lightest supersymmetric particle has to be stable. In this way at least one stable particle can be expected. It would still exist today and is hence a good candidate for Dark Matter. This should be the lightest particle, which is called LSP (lightest supersymmetric particle). In principle three particles predicted by Supersymmetry are candidates for Dark Matter as the LSP, since they are electrically and color neutral but interact gravitationally: The **sneutrino**, the **neutralino** and the **gravitino**. However, at least in the minimal supersymmetric standard model (MSSM) the sneutrino seems to be ruled out because of the width of the Z-boson decay at LEP [18].

---

[3]From the measured Z^0 decay it is known that up to $m_{Z^0}/2 \approx 45\,\text{GeV}/c^2$ only three neutrino families exist. However, if a heavier fourth neutrino exists, it is not expected to build up Dark Matter, since the abundance is estimated to be too small.

The neutralino would have properties like a heavy neutrino. Its lower mass limit is $m_\chi > 45\,\text{GeV}/c^2$ [13] and it is a superposition of the wino, the bino, and the uncharged higgsinos. The gravitino is also a candidate for Dark matter. However, since the gravitation is very weak and the gravitino interacts only gravitationally, it seems not to be directly detectable.

As the neutralino in Supersymmetry, many other theories predict a **weakly interacting massive particle** as candidate for Dark Matter. All these particles are summarized in the expression **WIMP**. It stands for a hypothetical particle which could solve the Dark Matter problem. In the following, WIMPs will be indicated with the neutralino symbol χ.

Up to now, out of all Dark Matter candidates, a WIMP seems to have the best chances of being detectable. Therefore, nearly all direct detection experiments try to realize this idea.

‖ **Most direct Dark Matter experiments aim to detect WIMPs.** ‖

In the next section the principle of a direct detection of WIMPs will be explained. From that different properties can be derived, which such Dark Matter experiments have to consider.

1.3 Direct Detection of Dark Matter in the Form of WIMPs

In section 1.1 we have seen many independent observations of Dark Matter. Also different theoretical explanations for these observations are already developed (section 1.2). Out of that, the direct detection possibility which seems to be most promising up to now is the search for WIMPs. Therefore, in this section, cornerstones for this kind of direct Dark Matter detection will be developed.

If the main contribution to Dark Matter are WIMPs, then these **weakly interacting massive particles** will be bound gravitationally in galaxies and galaxy clusters. The simplest assumption is a **spherical isothermal halo** around the luminous system which is filled with WIMPs. The dimension of such a halo has to exceed the luminous size by many times since the measured rotation curves show constant high velocities still far outside of the luminous masses. Due to the fact that WIMPs do not interact electro magnetically or strongly they can fly through the earth, sun, and galaxy nearly without any interaction. They interact even less than neutrinos and are simply bound gravitationally to the system.

However, from time to time, a weak interaction between a WIMP and the baryonic matter can occur. With typical velocities below the galactic escape velocity WIMPs should **scatter coherently** and **elastically** off nuclei. Via these scatterings energy can be transferred onto nuclei. The detection and identification of these nuclear recoils would be a **proof of the WIMP hypothesis**.[4]

---

[4]For simplification only the coherent WIMP scattering is considered in this work.

1.3.1 Interaction Rate

To see whether a direct detection of WIMPs is realistic the rate of expected WIMP interactions can be estimated:
In general the **detections R per time and detector target mass** M_T is given by the WIMP flux Φ_χ through the detector target, the number of target nuclei N_T, and their interaction cross section σ:

$$R = \frac{\Phi_\chi \cdot N_T \cdot \sigma}{M_T} = \frac{\Phi_\chi \cdot \sigma}{m_T} \tag{1.1}$$

m_T is the mass of one target nucleus.
In the following, the **WIMP flux** and the **cross section** will be estimated to get an idea about a possible rate of WIMP interactions:

- To evaluate roughly the **WIMP flux** through a detector target the local WIMP number density n_χ and their average velocity v relative to the target have to be known:

$$\Phi_\chi = n_\chi \cdot v$$

The local WIMP number density n_χ can be derived from the measured Milky Way rotation curve. As mentioned above, the local Dark Matter mass density ρ_χ at the position of the earth is estimated to be about $0.3 \, \text{GeV}/c^2 \, \text{cm}^3$. Thus, assuming the WIMP mass m_χ to be $100 \, \text{GeV}/c^2$, then about 3 000 WIMPs will be in each cubic meter:[5]

$$n_\chi = \frac{\rho_\chi}{m_\chi} \approx \frac{0.3 \, \text{GeV}/c^2 \, \text{cm}^3}{100 \, \text{GeV}/c^2} = 3\,000 \, \text{m}^{-3}$$

Assuming that WIMPs are at rest relative to the galactic frame and the detector target is moving through them with the velocity of the sun $v_\odot$. Then the average WIMP to target velocity is $v \approx v_\odot \approx 220 \, \text{km/s}$ [12] leading to a WIMP flux of

$$\Phi_\chi \approx 3\,000/\text{m}^3 \cdot 220 \, \text{km/s} \approx 65\,000 \, \text{cm}^{-2} \, \text{s}^{-1}$$

Considering now that the WIMPs have velocities not much faster than the sun velocity [12], the WIMP flux through the detector stays in the same order of magnitude.
If the assumptions made above are approximately correct, many ten thousands of WIMPs will fly through the size of a thumbnail every second.

- After the WIMP flux the next parameter for the rate estimation of equation (1.1) is the interaction **cross section** σ. The cross section depends on the two particles and the type of interaction. However, to be able to compare different Dark Matter experiments which have all different cross

---
[5] Assuming each kg of the human body has a volume of one liter, since humans consists mainly of water, then in each kg of mass there are, on average, three WIMPs. Thus, in the volume of a 50 kg human there are about 150 WIMPs.

sections due to the different target materials used, a normalized cross section σ_{norm} is usually introduced to separate the influence of the specific nuclei [19]:

$$\sigma_{norm} = \left(\frac{1+\frac{m_\chi}{m_T}}{1+\frac{m_\chi}{m_p}}\right)^2 \frac{\sigma}{A^2} \approx \left(\frac{m_T}{m_\chi}+1\right)^2 \frac{\sigma}{A^4} \quad (1.2)$$

$$\approx \begin{cases} \dfrac{\sigma}{A^4} & \text{for } m_T \ll m_\chi \\ \left(\dfrac{m_p}{m_\chi}\right)^2 \dfrac{\sigma}{A^2} & \text{for } m_T \gg m_\chi \end{cases}$$

A is the atomic mass number and m_p is the proton mass.

With this normalization the influence of the different target materials on the cross section for the coherent elastic scattering is taken into account. The normalized cross section is the value which different Dark Matter experiments can compare.

Assuming $m_T = m_\chi = 100\,\text{GeV}/c^2$, then A is about 107 and, using the largest possible cross section which is experimentally not excluded up to now $\sigma_{norm} < 10^{-7}$ pb:

$$\sigma \approx 3.2 \cdot 10^7 \cdot \sigma_{norm} \lesssim 3.2\,\text{pb}$$

If all these assumptions are roughly correct, the interaction rate of WIMPs with a detector target can be estimated. The detections of equation (1.1) per kilogram detector target and per year is given as:

$$\boxed{R \lesssim 50\,\frac{1}{\text{kg\,a}}}$$

Only a few ten WIMP interaction events can be expected for measuring one year with a detector of one kilogram of target mass. A **possible**, but very **difficult** challenge due to the usually large number of **background** events.

Only from this number of expected events it is clear that all possible external background events have to be suppressed as much as possible. This is usually done by using a laboratory deep underground and different shieldings. And it is clear that also internal background sources in the shieldings and detectors have to be reduced as much as possible. Nevertheless, the **passive background suppression** is usually by far not enough. An additional **active background suppression** is necessary. An identification of background events which could not be suppressed is needed. With such a low background experiment it could be possible to identify a WIMP signal.

> **An identification of WIMP interactions could be possible, but an extremely low background experiment is necessary for that.**

1.3.2 Deposited Energy

Beside the interaction rate the amount of energy E_{dep} transferred, and therefore deposited in the target is an important property for the direct WIMP detection. An energy transfer via elastic scattering depends mainly on the **masses** of the two particle and on their **velocities**.

The **WIMP mass** in different models is expected to be in the range of $45\,\text{GeV}/c^2$ to $3\,\text{TeV}/c^2$ [13][12]. To visualize possible WIMP masses one can compare them to known **baryonic masses**. One atomic mass unit u is about one GeV/c^2. Lead atoms as one of the heaviest, stable nuclei have a mass of about $195\,\text{GeV}/c^2$ and titanium for example has an atomic mass of about $45\,\text{GeV}/c^2$. The amount of transferred energy is maximal, when the masses of WIMP and target nucleus match.

The **velocities** can be estimated simply: The kinetic energy of a target nucleus is zero in the lab frame, whereas the average velocity of WIMPs can be approximated from the velocity distribution of the luminous matter in our galaxy. As above the average WIMP velocity is taken to be equal to the velocity of the sun $v_\odot = v_\chi = 220\,\text{km/s}$.

If WIMPs have a mass of $m_T = 100\,\text{GeV}/c^2$ then the kinetic energy of a WIMP is typically:

$$\boxed{E_\chi = \frac{1}{2} m_\chi v_\chi^2 \approx 50\,\text{keV}}$$

The energy transfer cannot be larger than the total kinetic energy of the WIMP. The maximum energy transfer in this case of $50\,\text{keV}$, is only possible for equal masses and head on collisions. Although WIMPs are maybe heavier and there exist higher WIMP velocities, in general the transferred energy will be lower due to the **scattering angle**.

> **The amount of energy which has to be detected is of the order of 10 keV.**

1.3.3 Energy Spectrum

If Dark Matter consists mainly of WIMPs, a direct detection has the two main properties shown above: A rough estimation of the transferred **energy** is of the order of $10\,\text{keV}$. The expected **rate** was evaluated to be not more than about a few ten events per target kilogram and year.[6] In this section, a more precise analysis of the expected **energy dependent rate** will be given.

If WIMPs interact with baryonic matter via coherent elastic scatterings, then these nuclei will receive energy. The amount of energy depends on different

---

[6] The cross section is assumed to be $\sigma_{norm} < 10^{-7}\,\text{pb}$ and the WIMP mass and the target nuclei mass to be $m_\chi = m_T = 100\,\text{GeV}/c^2$.

parameters. These are the two masses, the velocity of the incident particle, the scattering angle, and the form factor of the nucleus. For the interaction rate R, which depends on the deposited energy E_{dep}, additionally the target mass, the WIMP density and the cross section have to be taken into account.

In this section, first a simplified but already proper case of the **energy spectrum** will be presented. In this case one can see nicely from the formula how the dependencies on the various parameters are. Afterwards, a more realistic case will be shown. For this case the resulting formulas will be plotted to show finally, on the basis of the plot, that it does not differ much from the simplified case. The dependencies are nearly the same.

- In the simplified first case the WIMP velocity distribution is assumed to be Maxwellian without a cut off velocity. In reality the cut off velocity is the escape velocity of the galaxy. WIMPs faster than this velocity are not bound in the galaxy. Hence their density in the galaxy is negligible. However, this effect is neglected in the first simplified case. The motions of sun and earth are not taken into account, too. In this case the energy dependent interaction rate can be written as [12]:

$$\frac{dR}{dE_{dep}} = \frac{2\,\sigma_0\,\rho_\chi}{\pi\,v_\chi\,m_\chi\,\mu^2} \cdot F^2(E_{dep}) \cdot \exp\left(-\frac{2\,m_T\,E_{dep}}{\pi\,\mu^2\,v_\chi^2}\right) \quad (1.3)$$

μ is the reduced mass:

$$\mu := \frac{m_T \cdot m_\chi}{m_T + m_\chi}$$

σ_0 is the interaction cross section for zero momentum transfer and $F(E_{dep})$ is the form factor, which describes the mass distribution inside the target nuclei.

The energy dependent interaction rate per energy bin is the shape of the expected target nuclei recoil spectrum. It is given mainly by the **form factor** of the target nuclei and the characteristic **exponential decrease**.[7] For light elements the form factor is nearly constant $F^2(E_{dep}) \approx 1$; thus the recoil spectrum is roughly an exponential spectrum.

- Considering additionally the motions of sun and earth, the WIMP velocity has to be boosted into the earth's rest frame. For this calculation the velocity of the earth, which is equal to the detector target velocity, can be written as [12]:

$$v_T = v_\odot \left[1.05 + 0.07 \cos\left(\frac{2\,\pi\,(t - t_0)}{1\,\mathrm{yr}}\right)\right] \quad (1.4)$$

Here, $t_0 = \mathrm{June\,2nd} \pm 1.3\,\mathrm{days}$, the time when the earth has the highest relative velocity to the WIMP halo. $v_\odot \approx 220\,\mathrm{km/s}$ is the velocity of the

---
[7]The exponential decrease is given through the velocity distribution of the WIMPs.

1.3 Direct Detection of Dark Matter in the Form of WIMPs

sun in the galactic rest frame.

Taking into account that WIMPs with velocities higher than the escape velocity of the galaxy v_{esc} are not bound in the system, the Maxwellian WIMP velocity distribution has to be truncated. The escape velocity is expected to be $v_{esc} = 550 \pm 100\,\mathrm{km/s}$ [20] and is here taken as $650\,\mathrm{km/s}$. The recoil energy spectrum can then be written as [21]:[8]

$$\boxed{\frac{dR}{dE_{dep}} = \frac{\sigma_0\,\rho_\chi}{4\,v_T\,m_\chi\,\mu^2} \cdot F^2(E_{dep}) \cdot \frac{\frac{1}{2}\int_{\bar{v}_{min}-\bar{v}_T}^{\bar{v}_{min}+\bar{v}_T} e^{-\tau^2}\,d\tau - \bar{v}_T\,e^{-\bar{v}_{esc}^2}}{\int_0^{\bar{v}_{esc}} \tau^2 e^{-\tau^2}\,d\tau}} \qquad (1.5)$$

The velocities $\bar{v}_x$ are defined relative to the sun velocity $v_\odot$:

$$\bar{v}_x := \frac{v_x}{v_\odot}$$

v_{min} is defined as:

$$v_{min} = v_{min}(E_{dep}) := \sqrt{\frac{E_{dep}\,m_T}{2\,\mu^2}}$$

The form factor $F(E_{dep})$ describes the mass distribution in the target nucleus. Conventionally, the so-called Helm form factor, which was introduced by R. Helm [22], is used:

$$F(E_{dep}) = 3\frac{j_1(r_0 E_{dep})}{r_0 E_{dep}}\,e^{-\frac{s^2 E_{dep}^2}{2}}$$

$j_1(E_{dep}r_0)$ is the first Bessel function, which is responsible for the typical form factor shape, see below. s is the nucleus surface thickness. r_0 is the radius of the nucleus. Here, a parametrization of the nucleus is taken from [23].

The denominator in equation (1.5) $\int_0^{\bar{v}_{esc}} \tau^2 e^{-\tau^2}\,d\tau$ is the normalization factor of the Maxwellian velocity distribution.

Finally, the contribution $-\bar{v}_T\,e^{-\bar{v}_{esc}^2}$ in equation (1.5) reflects the WIMP cut-off velocity.

In figure 1.5 the energy dependent recoil spectrum of equation (1.5) is plotted for four different targets. The WIMP mass is assumed to be $m_\chi = 100\,\mathrm{GeV/c^2}$, the cross section to be $\sigma_{norm} = 10^{-7}\,\mathrm{pb}$. The different exposures are chosen so that the total interaction rate is in all cases identical. In the first plot the y-axis is linear, in the second logarithmic.

The linear plot shows nicely the **exponential decrease** of the expected WIMP interaction rate. In the logarithmic plot the influence of the **form**

---

[8] The equation is valid for $0.13\,m_T \lesssim m_\chi \lesssim 6.7\,m_T$.

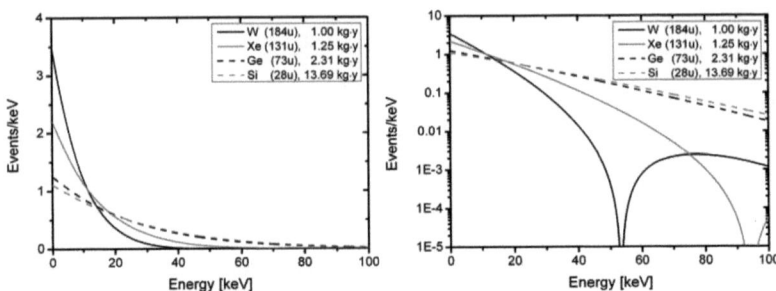

Figure 1.5: The energy spectra of equation (1.5) are plotted for four different targets. The WIMP mass is assumed to be $m_\chi = 100\,\text{GeV}/c^2$, the cross section to be $\sigma_{norm} = 10^{-7}$ pb. The exposures are chosen so that the total rate is always the same. In the first plot the y-axis is linear. Here the **exponential decrease** can be seen nicely. In the second plot the y-axis is logarithmic. In this plot, the influence of the **form factor**, which is dominated by the Bessel function, can be seen, in particular for heavy target nuclei.

factor is visible, especially for heavy target nuclei. The form factor suppresses for heavy target nuclei the larger energy transfers due to the Bessel function. It can be seen from the exposure that **heavier target nuclei result in higher expected interaction rates**.

From both descriptions, the simplified one with equation (1.3) and the more complex one with equation (1.5), the following can be seen:

> **The recoil energy spectrum is mainly a multiplication of an exponential decrease and the nuclear form factor.**

Dark Matter experiments entail the need of being able to detect a small number of nuclear recoils with energies of about 10 keV or even less. Additionally, the expected recoil spectrum is similar to energy spectra due to background events. Therefore, one of the main challenges is the possibility to avoid and identify background events. This will be discussed in chapter 2. How a WIMP signal could, nevertheless, be identified and distinguished from background signals will be discussed in the following section.

1.3.4 WIMP-Signature

For a Dark Matter experiment it is not enough to simply count events at about 10 keV, since also background events can be measured. Thus it is not possible to identify Dark Matter in the form of WIMPs only from an event rate at low energies. Furthermore, different background sources can induce events with an exponential energy spectrum, too, so that also a measured exponential energy

1.3 Direct Detection of Dark Matter in the Form of WIMPs

spectrum is not enough. A more characteristic signature is necessary to identify WIMPs clearly. How can such a signature look like? Up to now three possible signatures have been established:

- A convenient signature is the measurement and comparison of the interaction rates with at least two **different nuclei**. The two nuclei interaction rates can be measured within one or two different experiments. With the known sensitive energy range of the experiment, the expected rate can be determined with the help of figure 1.5. Since rates differ for different nuclei in a known way, see equation (1.2), Dark Matter could be identified by comparison of the two results.

- The second signature is more difficult since it has a drawback. It needs not only a very low background rate, but also a large number of detected WIMP interactions. It is based on the fact that the relative velocity of the earth to the WIMP halo depends on the seasons. In summer, the relative velocity is the highest and in winter the lowest, cp. equation (1.4). Therefore, in equation (1.5), the energy dependent rate changes. In figure 1.6 such a **summer-winter modulation** is shown. It can be seen that the effect is in the percent range.[9] Therefore large statistics is needed to confirm the modulation.

- The third signature could be realized on the basis of the identification of the **WIMP direction**. Since sun and WIMPs have similar velocities, the WIMP flux should come in average preferred from one direction through the detector. If a detector existed, which is able to identify the direction of nuclear recoils and thus the direction of the incident WIMP, WIMPs could be identified. The WIMP flux direction changes from day to night due to the earth's rotation. To check if the signal coming from one direction is extraterrestrial or earth-based, the detector could be turned. Unfortunately, a large enough detector with directional sensitivity could not be realized up to now.

In all these cases a statistically relevant number of WIMP scatterings is necessary. Such **statistics** can be achieved via a large **target mass**, a long **measurement duration**, and a low **WIMP identification threshold**. The WIMP identification threshold is defined as the transferred energy below which a WIMP signature cannot be observed any more in a specific experiment. These three parameters, which determine the sensitivity of a Dark Matter experiment, will be discussed in chapter 2.3.2.

Up to now, experiments are limited in all these three possibilities to increase the statistics by **background** events. Therefore, beside the WIMP scattering rate, the most critical point of a Dark Matter experiment is the background event rate.

---

[9]The modulation depends on the WIMP and target masses and is also sensitive to the halo model.

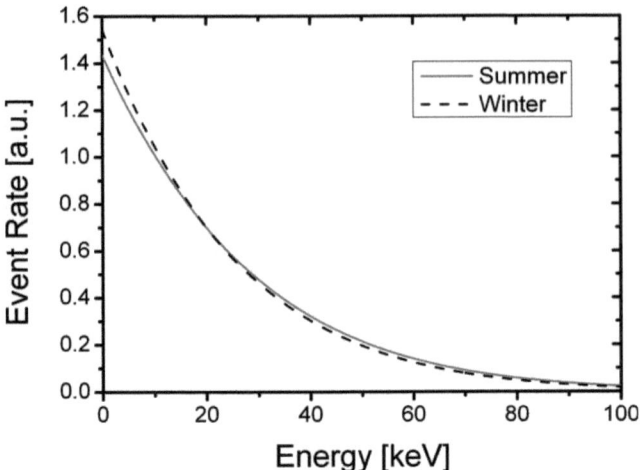

Figure 1.6: Energy spectra of $100\,\text{GeV}/c^2$ WIMPs detected with a germanium target ($A_{Ge} = 72.6\,\text{u}$) in summer and winter. The different shapes result from the change in the relative velocity between earth and WIMP halo. The relative velocity is the highest in summer. Therefore, slightly higher energy transfers are expected. In winter it is the other way round. The relative spectra are computed using formulas (1.4) and (1.5).

There will always be a rate of unidentified background events caused by one or more sources. This rate is a value per time and mass. If the number of background events is much higher than the WIMP signal, an increase of the detector **target mass** or a prolongation of the **measurement duration** cannot enable the WIMP detection. Therefore, an identification of the events or a suppression of the background source is necessary.

Also the **WIMP identification threshold** is limited due to background events. It is defined on the basis of these events. Below the WIMP identification threshold a WIMP signal cannot be measured, since the rate of background events is much higher than the WIMP interaction rate. A lowering of the WIMP identification threshold and therefore an improvement of the experiment's sensitivity can be achieved via a better identification of background events.

> ‖ **The statistics for a direct WIMP detection** ‖
> ‖ **is limited by background events.** ‖

In the next section, a limitation for the WIMP detection due to background events will be discussed for a concrete direct Dark Matter experiment. This will be the (**C**ryogenic **R**are **E**vent **S**earch with **S**uperconducting **T**hermometers) **CRESST-II Dark Matter** experiment.

Chapter 2

The CRESST-II Experiment

The CRESST-II experiment tries to detect Dark Matter in form of WIMPs. In this chapter CRESST-II will be introduced. First, its **setup** will be presented in section 2.1. Afterwards, the mode of **operation** will be explained in section 2.2. Section 2.3 comes back to the cornerstones of the direct WIMP detection of section 1.3.4, which determine the **sensitivity** and which will be discussed for the CRESST-II experiment. Finally, out of this discussion the **motivation** and goal of this work will be presented in section 2.4.

2.1 CRESST-II Setup

As all direct Dark Matter search experiments trying to detect WIMPs, two major challenges exist for CRESST-II: Firstly, to detect low energetic single interactions with a rate of not more than about ten interactions per kilogram of target material and year, and secondly to reduce and identify background events.
The principle of **background reduction and identification** in CRESST-II will be presented in section 2.1.1. The **detector modules**, which are able to detect single interactions with energy transfers of even less than one keV, will be introduced in section 2.1.2.

In figure 2.1 the setup of CRESST-II is shown. The detector modules can be seen in central position. They are surrounded by different layers of shielding. On the top, the cryostat can be seen, which provides the required low temperature conditions for the detector modules.

2.1.1 Background Reduction and Identification

As mentioned in the first chapter, background events are the most crucial point for a direct Dark Matter search experiment. Reasons for this are that the background itself is unknown, which prevents a simple subtraction, and that the scattering rate of WIMPs is very low. Therefore most effort of Dark Matter experiments is done in order to reduce and identify background events.
Reduction can be achieved via **removing background sources** and via the installation of **shieldings**, which protect the detectors. The **identification**

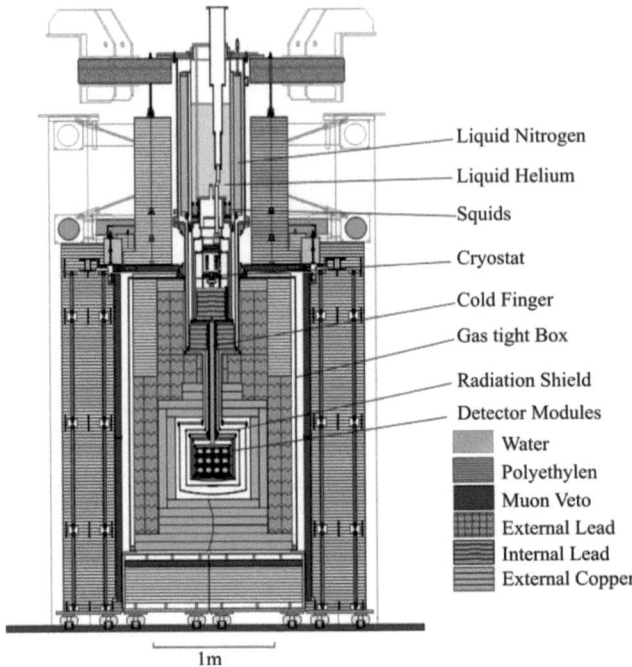

Figure 2.1: In this schematic drawing the CRESST-II setup can be seen. Different layers of shielding enclose the very center. These are, from the outside, water, polyethylene, the muon veto, lead, and copper. On the top of the experiment the longish cryostat can be seen. It provides the low temperature, which is required for detector operation. The cryostat is thermally connected via a copper post, called cold finger, to the detector modules. They are located in the very center enclosed by the shieldings.

of background events is based on the differences in the interaction between WIMPs and background.

> **Reduction of background can be achieved by removing sources, suppressing background interactions, and identifying background events.**

In the following these three possibilities will be explained in more detail.

Reduction of Background Sources

The best way to prevent background events is to remove the source. Removable background sources are in the material **volumes**, on their **surfaces** or in the **surrounding gas**.

2.1 CRESST-II Setup

Volume: To prevent sources in the **volume** special **radio-pure materials** are used. The closer the material is to the detector, the less radioactive it should be. For the shielding around the detector, for example, roman lead is preferred due to the large proton number and the low intrinsic activity. Generally, in lead the intrinsic activity is dominated by ^{210}Pb. In case of roman lead, on the other hand, the material has not been in contact with the other isotopes of the radium series (also called uranium series) for many years. Hence, a further ^{210}Pb production is suppressed efficiently and the existing ^{210}Pb nuclei isotopes have decayed with a life time of 138 d. Another possibility for a reduced intrinsic volume activity of a material is the underground storage. This is efficient in case of dominant activation due to the cosmic radiation, as it is the case, for example, for copper. For this reason in CRESST-II copper is stored in an underground lab.

Surfaces: Typically background sources are most often found on **surfaces**. Therefore, if possible, all surfaces should be **cleaned mechanically and chemically**. The material should then not be touched anymore with other radioactive matter. Human sweat, for example, contains the beta emitter ^{40}K. Hence, bare hands should not touch inner materials. In general the storage of the materials in a **clean room** should be favored.

Surrounding Gas: Also in the **surrounding gas** there are dangerous background sources. Rock, for example, emits ^{222}Rn, which is radioactive. Therefore, CRESST-II uses an air tight box which surrounds the detectors. An overpressure of radio-pure nitrogen is maintained in this box to prevent radon close to the detectors. Materials should hence be surrounded by **non-radioactive gases** under overpressure.

Passive Background Reduction

Since not all background sources can be removed, the next step for the reduction of unwanted background events is the protection of the detectors by shieldings. For almost each kind of **source**, there exists a special shielding. From outside these are in CRESST-II the following:

First, to reduce background events induced by **cosmic radiation** or their secondary products, Dark Matter experiments are usually located in a deep underground lab. The CRESST-II experiment is located in the Laboratori Nazionali del Gran Sasso (LNGS), a lab in the center of Italy placed in a highway tunnel of the Gran Sasso mountains. The protecting rock there is at least 1 400 m thick, which is equivalent to a water depth of about 3 150 m. In that deepness cosmic radiation induced events are almost gone. The amount of **muons** is reduced by about six orders of magnitude to a flux of about $1/\text{h}\,\text{m}^2$ [24]. Not suppressed are only neutrinos, whose energy transfers onto nuclei should be of less than 1 keV [25].

The second protection reduces **electro-magnetic interferences**. In CRESST-II highly sensitive electronics is used which can be influenced by electro-magnetic interferences. For this reason a protection by a faraday cage is necessary which encloses and protects the experimental setup.

In a Dark Matter experiment single interactions have to be detected. Due to this sensitivity external **vibrations** can generate background events, where no real particle interaction took place. Therefore, the whole inner part of the experimental setup is positioned on air dampers. Additionally the detectors in the very center of the setup are mounted on springs to reduce the influence of vibrations even further.

The next step is the protection of the experimental setup against gamma radiation and neutrons. For this various layers of shielding are used: The outer-most of these shieldings is polyethylene of a thickness of about 50 cm and a mass of about 10 t. Polyethylene consists of carbon and hydrogen. Since hydrogen atoms have a similar mass as **neutrons** the energy transfer in a scattering is maximized. Hence, this shielding reduces the kinetic energy of incident neutrons effectively. As next inner shielding a 20 cm thick lead shield of 24 t is installed. Lead has a high proton number Z. With such a shield **gamma radiation** can be reduced effectively. The innermost shielding is a 14 cm thick and 10 t heavy copper shield. The advantage of copper is the low intrinsic activity. In this way it can shield lead gammas without introducing too much further radiation due to its low intrinsic activity.

Active Background Reduction

Although background sources can be removed and or shielded, there will always remain some background events. Usually the sources of this background are mostly located in the detectors themselves, and it is thus difficult to avoid them. Since they cannot be avoided the only possibility is to discriminate them from WIMP interactions. This procedure is named active background reduction. The basis of the active background reduction is a **difference between background and WIMP interactions.**

Background interactions differ from WIMP interactions and can therefore be distinguished. For this kind of differentiation two properties of the WIMP interaction are important: Firstly, the **interaction rate is extremely low** so that WIMP interactions are single interactions. The second important property is that WIMPs will transfer part of their **energy onto the heaviest nuclei** and not onto electrons and, if present, lighter nuclei. This is due to the assumed coherent WIMP interaction, which favors heavy nuclei.

The second property can be used, since events can be distinguished depending on which component energy is deposited. If the energy is <u>not</u> transferred onto the heaviest nucleus but onto electrons or lighter nuclei, the event is most likely background and can be rejected.

Neutrons deposit most energy on the lightest nuclei due to the mass ratio. The energy transfer onto the heavier nuclei is much smaller and usually below the detection energy threshold. **Alpha particles** deposit their energy partially onto light nuclei and partially onto electrons due to the electro-magnetic interaction. **Beta** and **gamma** radiation transfer energy onto electrons.

In all these cases the energy is not transferred onto the heaviest nuclei as it is for **WIMP** events. Thus, in principle such background events can be identified.

The first mentioned property of WIMP interactions, namely the low WIMP interaction rate, can also be used as background identification. If two events are measured coincidently it is most likely that they have the same cause. WIMPs, on the other hand, do interact extremely seldom so that **coincident** events, where at least one event is caused by a WIMP, is almost excluded. This is the reason why coincident events can be identified as background events. A coincidence can occur within two different detectors or one detector and an additional, external veto detector, which is installed especially for this reason.

For example, **muons** can be identified with a muon veto. In this case, the muon veto is the second external detector. If an event is measured in one of the detectors and coincidently the muon veto detects a muon, the detector event is most likely muon induced and will be ignored. Therefore, in CRESST-II a muon veto is surrounding the detectors, cp. figure 2.1.

Another example for an external veto detection are alpha decays, where the alpha particle is not detected, but the **daughter nucleus** is. The alpha decay is a two body decay. Hence, the alpha particle and the daughter nucleus have opposite directions. If the nuclear parent is on the detector surface, it can happen that the alpha particle is not or only partially detected, whereas the daughter nucleus is. The alpha decay cannot be identified in this case, since the alpha particle is not measured. Since these daughter nuclei are heavy they can look like recoil nuclei induced by WIMP interactions. A similar situation is given when the nuclear parent is on the surrounding surface. In this case it can also happen, that only the daughter nucleus is detected.

However, an additional signal can be produced, if the alpha particle can be detected. In CRESST-II a scintillating foil, which absorbs these alpha particles, is surrounding the detector for this reason. Via this absorption the alpha particle produces additional scintillating light, which can be detected and used as a veto signal.

In figure 2.2 the schematic drawing of such a decay is shown for the well known case of ^{210}Po $\longrightarrow$ ^{206}Pb $+ \alpha$. ^{210}Po is a daughter nucleus of the noble gas ^{222}Rn, which can be released from the surrounding rock into the air. Radon can move and get in contact with all different kind of surfaces.

In the figure also the detected energy ranges are shown, for each of the two possible cases: The nuclear parent can be in the detector volume close to the surface or on the surface of the surrounding material. The kinetic energy of the daughter nucleus after the decay is 103 keV. Hence, the detected energy is in the first case at least and in the second case not more than 103 keV. This depends on how much of the alpha particle energy is detected in addition or which fraction of its energy the daughter nucleus lost in the surrounding surface, respectively. Since WIMPs should transfer energies of the order of 10 keV, the alpha decays on the surfaces of the surrounding material are most disturbing. Therefore the detection of the alpha particle is essential to identify these kind of events.

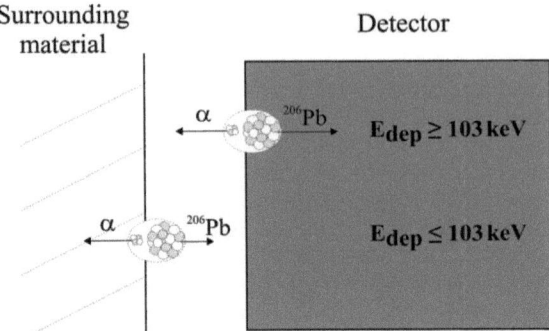

Figure 2.2: ^{210}Po contaminations can be on all surfaces. In the case of a CRESST-II detector module, this can be the target or the surrounding surface. If only the lead daughter nucleus is detected by the detector, it can look like a WIMP interaction. Since WIMP interactions have energies of about 10 keV, in particular the second case can be a disturbing background. In the first case at least an energy of 103 keV from the daughter nucleus is detected. Additionally, part of the alpha energy can be detected, i.e. the total amount of detected energy is at least 103 keV. On the other hand, in the second case the alpha particle does not deposit energy in the detector. Moreover, the daughter nucleus can lose a part of its energy in the surrounding surface, so that the total amount of detected energy in the crystal can be in the range of a typical WIMP interaction. This background is vetoed in CRESST-II by detecting the alpha particle as well.

Beside muon interactions and alpha decays, low energetic **neutrons** can be identified via coincident interactions. Their typical mean free path is in the range of about 10 cm. Therefore, two successive interactions in neighboring detectors are not unlikely. For **gamma** radiation of the order of 100 keV it is the same case. A gamma can transfer a part of its energy onto an electron of the target and escape the detector. A second detector target can absorb this escaping gamma.[1]

Also **electro-magnetic interferences** and **vibrations** occur most likely on more than one detector. Therefore, coincident event detection identifies these events as background.

2.1.2 CRESST-II Detector Modules

In the previous section the different kinds of shieldings, which can be seen in figure 2.1, have been discussed. In this section the **detector modules** in the very center of the figure, which are enclosed by the shieldings, will be introduced.

As seen in section 1.3, a WIMP detector has to fulfill mainly the following two requirements:

---

[1] In practice these interactions are identified by the way explained first.

2.1 CRESST-II Setup

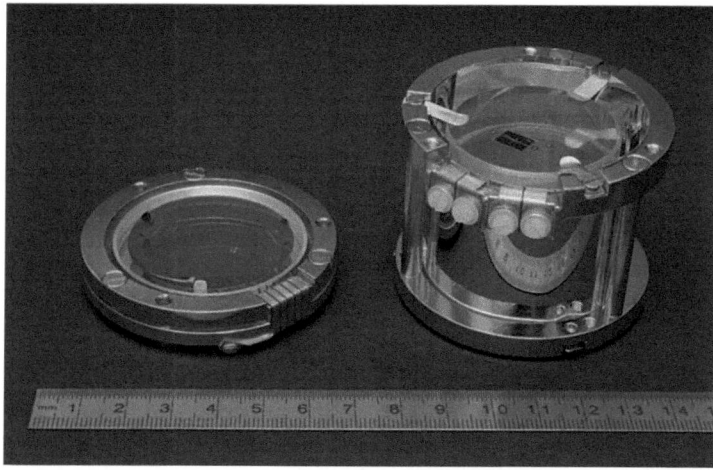

Figure 2.3: In this picture an opened detector module can be seen. On the left hand side the (brownish) light absorber wafer is placed. On the right hand side the scintillating CaWO$_4$ target crystal can be seen. Both are held by clamps in a light reflecting housing, framed by a copper holder.

- The detectors have to be able to detect energies of about 10 keV.
- The detectors have to be sensitive for a detection rate of not more than a few times ten interactions per kilogram detector target and year, i.e. they have to be able to actively identify background events, which typically occur at a much higher rate.

These two requirements are fulfilled by the CRESST-II detector modules. The setup of these modules will be introduced next.

In the CRESST-II experiment the detector consists of detector modules which themselves consist of two parts. In figure 2.3 the two parts of an opened detector module can be seen.

The smaller part, on the left hand side, is mainly a brownish, 40 mm in diameter and 0.46 mm thick sapphire (Al$_2$O$_3$) wafer. On the bottom side of the wafer a 1 μm silicon (Si) absorption layer is grown on, which gives the color. On the upper side of the wafer, there is a thermometer structure of about 1.5 mm × 2 mm in size. The weight of the wafer is 2.3 g and it is held by three clamps, which can also be seen. This smaller part of the detector module is called **light detector**. The light detector is finally positioned on the lower front side of the cylindrical scintillating calcium tungstate (CaWO$_4$) crystal, which can be seen on the right hand side. This crystal is the detector target. On the upper front side of the crystal a second thermometer is placed. Its size is 6 mm × 8 mm. The crystal itself is 40 mm high, 40 mm in diameter, and weighs about 300 g. In total, twelve

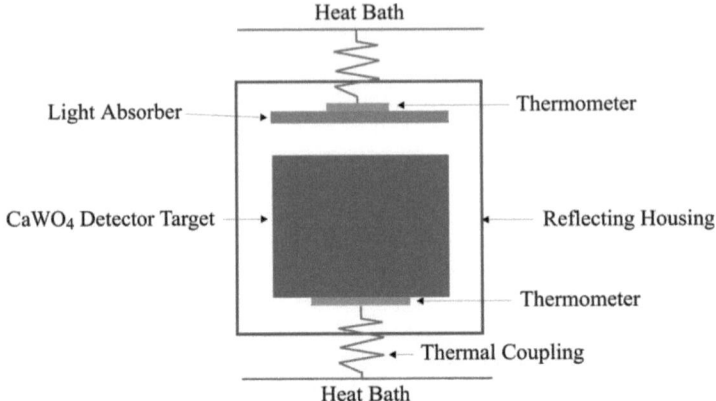

Figure 2.4: This schematic drawing shows a closed CRESST-II detector module. The main volume is filled by the target crystal. Its temperature is measured by a thermometer and it is thermally coupled to the heat bath. The light absorber's temperature is measured by a second thermometer and it is also thermally coupled to the heat bath. The detector setup is enclosed in a light reflecting housing.

clamps hold the crystal. Out of those, only three top clamps and one bottom clamp can be seen. This larger part of the detector module is called **phonon detector**.

Both detectors are surrounded by a **light reflecting housing**. Therefore the upper face of the phonon detector is finally covered by a reflecting end cap, which cannot be seen in the picture.

2.2 CRESST-II Operation

In the previous section the setup of the shieldings and the detector modules of the CRESST-II experiment have been presented. In this section the functional principle of these detector modules will be explained.

2.2.1 Signal Measurement

In figure 2.4 a schematic drawing of a closed detector module is shown. The main volume of a detector module is filled by a $CaWO_4$ crystal. This crystal is the target, which is sensitive for energy depositions. As required generally in section 2.1.2, a WIMP detector target has to be able to detect energy transfers of about 10 keV. To fulfill this, in CRESST-II the crystals are operated as **cyrodetectors** with **transition edge sensors** (TES).

A **cryodetector** is a detector which is operated at low temperatures. One advantage of a low temperature experiment is that heat capacities of the detectors

2.2 CRESST-II Operation

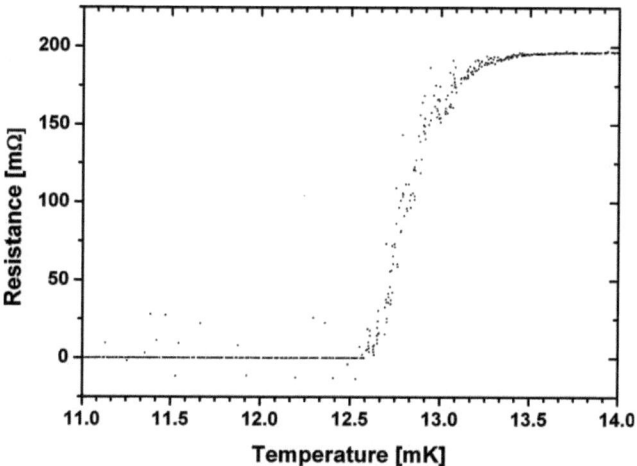

Figure 2.5: A typical temperature dependent resistance measurement of a transition edge sensor (TES). To use it as a thermometer it is thermally stabilized in the transition between the super-conducting and the normal-conducting state. A relatively small temperature rise results in a relatively large resistance change. The TES is hence a highly sensitive thermometer in the small temperature range of the transition.

can be reduced by orders of magnitudes. In CRESST-II the operation temperature is about $10 - 20 \, \text{mK}$. These temperatures are provided by a cryostat, which can be seen on top of figure 2.1. At such a low temperature, the heat capacity of non-metals is reduced by about 13 orders of magnitude compared to the room temperature value. The temperature rise, which is detected after an energy deposition, is inversely proportional to the heat capacity.

However, the temperature rise which has to be detected is still in the µK range due to the small amount of deposited energy and the macroscopic target volume. This is the reason why an extremely sensitive thermometer has to be used. The thermometer in case of CRESST-II is a **transition edge sensor** (TES). The main property of a TES is an extreme temperature sensitivity in a very small temperature range. A TES can be seen in figure 2.3 on the top face of the crystal. Physically, a TES is a metal film which is thermally stabilized in the transition between the superconducting and the normal conducting state. In this temperature region the thermometer resistance is sensitive to temperature changes of a fraction of a millikelvin. A typical temperature dependence of the TES resistance can be seen in figure 2.5.

As a simplified understanding of the detector functionality can be said:[2] The strongly reduced heat capacity at **low temperatures** and the extremely sensi-

---
[2] In reality the processes going on in the detector itself are more sophisticated. Chapter 5 will concentrate on that.

Figure 2.6: Two scintillating $CaWO_4$ crystals which are excited by ultraviolet light. The (bluish) scintillation light of the crystals can be seen. Location and direction of the light emission can be influenced by the shape and the roughness of the crystal surface, see chapter 4.

tive thermometer provided by a **transition edge sensor** allow to **detect energy depositions of even less than 1 keV** onto a macroscopic target by its temperature rise.

The **basic function** of a CRESST-II detector can be summarized as follows: After an energy deposition the target warms up. Across the thermal coupling to the heat bath the deposited energy is taken out of the target and the target cools back again. The temperature rise and cooling down of the target is detected by a thermometer. The comparison of the detected temperature rise and the temperature rise from a known energy deposition, gives the possibility to determine the amount of energy deposited.

2.2.2 Signal Differentiation

One important requirement for a WIMP detector is the ability to **identify background events**. For this reason, in the CRESST-II experiment a second signal is measured. This signal is based on the **scintillation light** which is emitted by the $CaWO_4$ target crystal.

A picture of scintillating $CaWO_4$ crystals can be seen in figure 2.6. Here the scintillation light is produced by exciting the crystals with ultraviolet light; generally it can be produced by an energy deposition in the crystal. After such an energy deposition where a fraction of the energy is transformed into scintillation light, a part of the emitted light can be absorbed by the light absorber of the detector module. A light absorber can be seen in figure 2.3. It is placed on one front side of the scintillating crystal, see figure 2.4. Crystal and absorber are enclosed by

2.2 CRESST-II Operation

a light reflecting housing to increase the amount of light collected by the light absorber. The absorber is warmed up due to the light absorption. This temperature increase is measured via a second thermometer, which is a TES, too. This signal forms the second measured channel of the CRESST-II experiment.

The physical basis of these two channels is, in general, that the energy deposited by WIMPs or background in materials can be transformed into phonons, scintillation light, ionization, and vacancy production. The splitting-up into these different channels depends on the material and on the type of interacting particle. However, one or more of these components can be detected to get information about the energy deposition. The CRESST-II experiment detects, as seen, phonons and scintillation light. Therefore, after an energy deposition in the detector target both thermometers measure a temperature rise. The first one measures the temperature rise of the target and the second one the temperature rise of the light absorber. After the temperature rises, target and light absorber temperatures relax back to the bath temperature via their thermal couplings. The two thermometers are read out by SQUID-based (**s**uperconducting **qu**antum **i**nterference **d**evice) electronics. These two signals are called light and phonon signal. The respective measured channels are therefore called light and phonon channel.

> The phonon channel consists of
> the target crystal and its thermometer.

> The light channel consists of
> the target crystal, the reflecting housing,
> the light absorber, and its thermometer.

For each energy transfer onto the detector target both channels are read out and measure a signal. In the following will be described, how an identification of background events is then possible.

As already mentioned in section 2.1.1, the active discrimination of background events is based on the difference of background interactions compared to WIMP interactions. Beside the low interaction rate, which indicates coincident events to be background, the energy transfer onto the heaviest target component is expected to be typical for WIMP interactions. Since energy transfers onto different components of the target crystal can be distinguished, an event by event discrimination is possible. The reason for it is that the ratio of the deposited energies, which are transformed into phonons and scintillation light, depends on the **mass** of the incident particle and the fact if it takes part in the **electro-magnetic interaction**.

Since WIMPs are expected to be heavy and do not interact electro-magnetically, they transfer energy onto the heaviest nuclei. Whereas alphas, betas, gammas, and neutrons transfer detectable energies onto lighter components of the $CaWO_4$

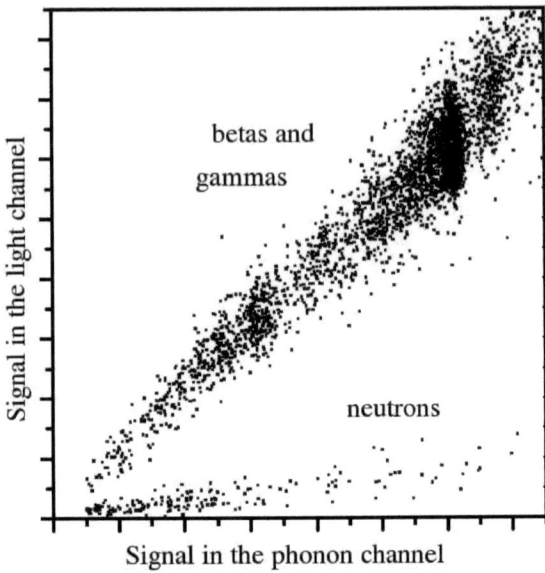

Figure 2.7: A CaWO$_4$ target crystal was irradiated first by beta- and gamma-radiation and afterwards by a neutron source [26]. The coincidently measured pulse heights of the light and the phonon channel are shown. It can be seen, that the ration of the light to the phonon signal depends on the incident particle. In this way an identification of the incident particles can be realized by the help of the second measured channel.

target crystal. As will be seen in chapter 3, these components result in a higher amount of produced scintillation light. For them the fraction of energy which is detected in the light channel is relatively large, whereas the fraction of energy in the light channel will be relatively small for WIMPs. Hence the coincidently measured ratio of the light to the phonon channel after an energy deposition allows an identification of background events.

> **The ratio of light to phonon channel signal depends on the incident particle type.**

This ratio is the basis of the active background discrimination. The principle proof for CaWO$_4$ was done first in CRESST-II. The coincidently detected signals of the two channels of this measurement can be seen in figure 2.7. In this measurement, a CaWO$_4$ crystal was irradiated by beta and gamma radiation and afterwards with neutrons. The coincidently measured two signal channels are plotted in the **light-phonon-plane**. The two different event bands are marked in the picture. It can clearly be seen that the ratio of the two measured channels depends on the incident particle type.

2.3 CRESST-II Sensitivity

The last section has shown how low energy depositions onto a macroscopic target are measured in CRESST-II and how background events can be identified. In the following section 2.3.1, it will be described where in the light-phonon-plane the different events, especially WIMP events, are expected. The comparison with the area where background events are expected shows how the sensitivity of CRESST-II is determined. Afterwards, in section 2.3.2, it will be clarified which parameters are crucial for the CRESST-II sensitivity.

2.3.1 Light-Phonon-Plane

In this section the essential point of discrimination of different particle types will be discussed in more detail. Furthermore, the energy range in the light-phonon-plane which is relevant for a WIMP detection will be discussed.

Assuming an incident particle deposits an energy E_{dep} in a $CaWO_4$ target crystal. Then light and phonon channel measure a signal. The ratio of the two signal amplitudes of these two channels depends on the incident particle. How is the deposited energy E_{dep} split-up into the two measured channels? What determines the splitting?

It is clear that the splitting cannot be determined by the weak interaction, since this interaction is overlapped and dominated by the electro-magnetically interaction. **The splitting is done via the electro-magnetic interaction**, only charged particles can determine the splitting. In other words: For each type of charged particle a specific splitting exists. **For each type of charged particle a band shows up in the light-phonon-plane.** For uncharged particles, as neutrons or WIMPs, it is of importance on which type of charged particle energy is transferred.

In the following the relevant bands of the CRESST-II experiment will be presented. Afterwards, properties of these bands in the light-phonon-plane will be determined. With the help of this information a schematic drawing of the light-phonon-plane with the most important bands will be developed. On the basis of this drawing the key points of the CRESST-II Dark Matter detection and its sensitivity can be seen.

- **Alpha particles** (α) are charged, therefore these events have a specific splitting ratio of the deposited energy into the two measured channels - the light and the phonon channel. For this reason alpha-radiation shows up in an own band.

- **Beta radiation** (e) are charged particles, too. They appear within their band in the light-phonon-plane.

- **Gamma radiation** (γ) is very similar to beta radiation, since in the case of interest, $E_{dep} < 100\,\text{keV}$, the photoelectric effect is absolutely dominant.

A small difference, however, shows up, since the photoelectric effect releases most likely an inner electron of an atom. This electron has a kinetic energy reduced by the binding energy. The created hole is successively filled up, the released energy releases further electrons, so that finally a few electrons with smaller energies are created compared to a single electron of the same total energy in case of beta radiation. The splitting ratio of the deposited energy can be slightly energy dependent. Thus, for a single electron the splitting can be different compared to the splitting of a few electrons with the same total energy. For this reason the band due to gamma radiation is not exactly the beta band. However, in the following this effect will be neglected, so that the two bands are handled as one.

- **Neutrons** transfer energy by strong interaction. Therefore, it is of interest onto which particle they transfer energy. Due to the mass ratio neutrons transfer the largest amount of energy onto **oxygen (O)**. This oxygen recoil itself loses its energy in the crystal via the electro-magnetic interaction. For this reason there exists a specific energy splitting for oxygen, which shows up in form of a band.

- Alpha decays, in general, create recoiling daughter nuclei with certain kinetic energies. Without detecting the alpha particle and in case of absorption of the nuclear recoil in the target, the **heavy nuclei** (e.g. **Pb**) show up in additional bands in the light-phonon-plane. These events can usually not be distinguished from WIMP events. Therefore, the detection of the alpha particle is important.

- **WIMPs**, on the other hand, are expected to induce events in the **tungsten (W)** band, since they should interact mainly with the heaviest nuclei of the target. Therefore, **the tungsten band is the band of interest**.

- The last component of the $CaWO_4$ target is **calcium (Ca)**. It is not expected that any type of incident particle dominantly transfers detectable energies onto these nuclei. Neutrons show mainly up in the oxygen, and WIMPs in the tungsten band. However, although less events are expected in this band compared to the other bands, a band for calcium exists in the light-phonon-plane.

- In addition to these bands, two other bands can show up in the light-phonon-plane. These are events, where only one channel detects a signal. This can happen if, for example, energy is transferred onto the light detector directly. In this case no energy is transferred onto the target. Only the light detector measures a signal. Such an event is called **only light** event (**OL**). On the other hand, there exist **only phonon** events (**OP**), too. These are events, where only the target measures a signal. Formation of cracks in the crystal due to stress in the lattice can be a reason for such events. In the light-phonon-plane these two bands are placed on the two axis.

2.3 CRESST-II Sensitivity

These are the most important bands of the CRESST-II experiment. To be able to make a schematic drawing of these different bands in the light-phonon-plane, different properties of all these bands and the plane itself are listed next. With this drawing a helpful overview of the coincidently measured signals of CRESST-II is given.

- From figure 2.7 it can be expected that the slopes of the bands have no or only a small energy dependence. This has been verified by the CRESST-II experiment. The band middle lines are to first order **lines through the origin**.

- The ratio of the coincidently measured light channel to phonon channel signal amplitude is fixed for each band. It is typical for the respective particle and corresponds to the **slope** of an event band in the light-phonon-plane. The slopes are indicated by the so-called **quenching factors**. They are defined as the reduction of the slope compared to the slope of the gamma band. In other words, the quenching factor is the reduction of the measured light of a particle compared to the light of a gamma quantum of same energy.[3] The physical reason for this light quenching, i.e. the different splitting, will be discussed in detail in chapter 3.

- One important property of these bands is their width. The **width of a band** is given by the energy dependent **resolutions** of the two measured signal channels. The energy resolutions will be discussed in chapter 4.

- The deposited energy E_{dep} is split up depending on the particle. For this reason the **calibration** of a channel is particle dependent. Since most of the deposited energy is in any case transformed into phonons, the phonon channel calibration, however, is nearly particle independent. The absolute energy variation can be neglected. For the light channel, on the other hand, the absolute detected energy varies by about a factor of 30 depending on the particle type. The calibration is strongly particle dependent. For this reason one calibration has to be chosen. In case of CRESST-II, this is the gamma calibration. Since gamma-radiation transfers energy onto electrons and for historical reasons[4] the choice of the gamma calibration for the light channel is marked at the energy unit by using *electron equivalent*: keV$_{ee}$

With the help of the above information a schematic drawing of the bands in the light-phonon-plane is possible. The bands represent the area in the plane where measured signals will show up, if the respective particle deposits energy in the target.

---

[3] The absolute signal in the phonon channel is nearly particle independent since in all cases almost the deposited energy is transformed into phonons.
[4] The differences of the gamma and the electron band are minimal and were unknown at the time of introducing keV$_{ee}$.

Assuming an infinitely good energy resolution of both measured channels, i.e. the width of the band is zero, then the electron (e), the alpha (α), the oxygen (O), the calcium (Ca), the tungsten (W), the only light (OL), and the only phonon (OP) bands show up as shown on the top left picture of figure 2.8. The slope of a band is a physical property, which will be discussed in chapter 3. The light reduction, i.e. the reduction of the slope compared to the electron band which is described by the quenching factor is assumed to be 1/5 for alpha particles, 1/10 for oxygen, 1/20 for calcium, and 1/30 for tungsten.[5] As seen in section 1.3.3 energy transfers of WIMPs of more than 40 keV are unlikely. Therefore, only energies below 50 keV are shown in the figure.

Not to overfill the following plots only the three most important bands will be taken into account: The only light band (OL) can be neglected, since it is far away from the tungsten band in the light-phonon-plane. Therefore, these only light events can be distinguished easily from tungsten recoil events, which can be WIMP induced. The same situation is given for alpha events. The reason here is that they appear at energies of a few MeV, which is about three orders of magnitude higher than the energy depositions expected from WIMPs. For the calcium nuclei no source is known, which dominantly transfers energy onto them. Therefore, no events are expected to be dominantly in the calcium band. Only phonon events (OP) appeared in CRESST-I as well as in CRESST-II, when the crystal was held too tightly, so that too much stress was induced in the lattice of the target crystal which can release in form of microscopic cracks [27]. At the moment, this seems not to be the case for CRESST-II. For these reasons the bands are skipped in the following.
Beside the tungsten band (W), the electron (e), and the oxygen band (O) are left over. These three bands are shown on the top right picture of figure 2.8. In this plot, compared to the left one, the y-axis is changed from the light channel signal to the ratio of this signal to the phonon channel signal. This way bands with constant slopes are transformed into horizontal bands. This kind of drawing has the advantage to show the low energetic area (< 10 keV) in detail as well as the higher energetic area (> 40 keV) in one and the same picture.

In reality, all bands have finite, energy dependent widths. The width of a band is given by the energy resolutions of the two measured channels. The two middle pictures of figure 2.8 show the three bands of interest with typical widths of CRESST-II taken from [28].[6]
The energy resolution of the phonon channel is about one order of magnitude better than the one of the light channel for same energies. Therefore, their influence on the width of the band is neglected.
The width of the tungsten band is chosen so that 80 % of the events appear in

---

[5]The simplified quenching factors used here are rounded values of quenching factor measurements, cp. chapter 3.
[6]The 1σ energy resolution taken as $\Delta E = \sqrt{1.44\,\text{keV}^2 + 0.33\,\text{keV} \cdot E + 0.01 \cdot E^2}$ is a typical derived value of CRESST-II light channels.

2.3 CRESST-II Sensitivity

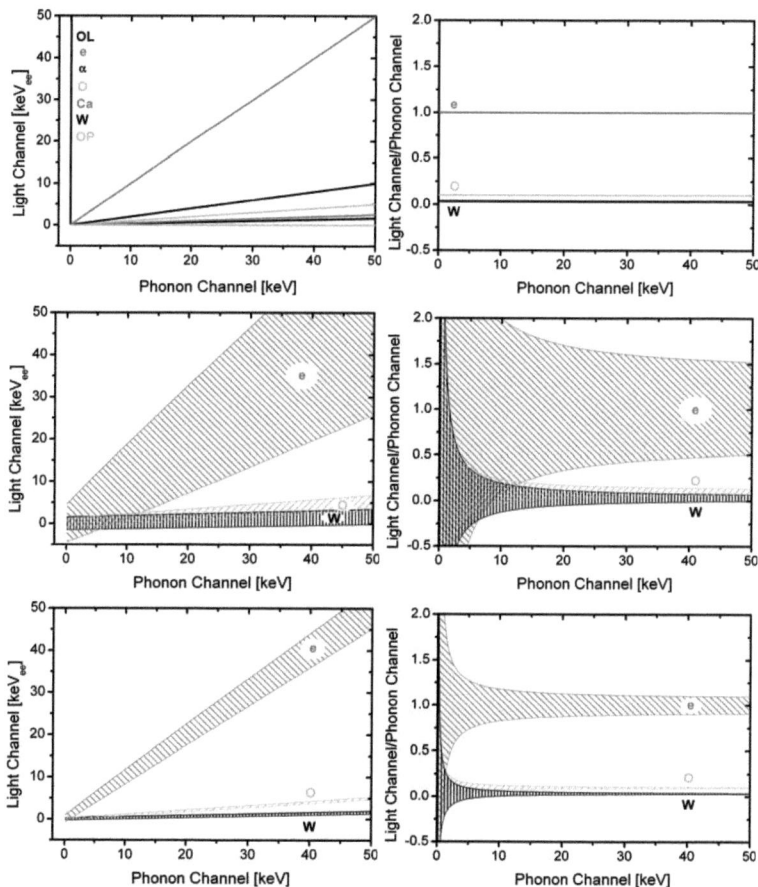

Figure 2.8: The three pictures on the left hand side show the bands in the light-phonon-plane. For each type of particle one band shows up. The three pictures on the right hand side are identical to the three on the left hand side, apart from the fact that the y-axis is instead of the light channel the ratio of this channel to the phonon channel. Therefore, bands with fixed slope on the left hand side change to horizontal bands on the right hand side. The first picture shows the electron (e), the alpha (α), the oxygen (O), the calcium (Ca), the tungsten (W), the only light (OL), and the only phonon (PO) bands. The other pictures show only the three most important bands for better clearness. The energy resolutions of the two measured channels of the first two pictures are assumed to be infinitely good, for the two middle pictures typical values of CRESST-II are taken for the energy resolutions. For the lower two pictures the energy resolution of the light channel is improved by a factor of five. Depending on the energy resolution the events are more concentrated or more spread.

the band area for each energy (1.28 σ energy resolution of the light channel). The tungsten band is the area where WIMP events are expected.

|| **WIMP events show up in the tungsten band of the light-phonon-plane.** ||

This is the sensitive area of the CRESST-II experiment. In all other bands only background events show up.

The width of the electron band is selected in such a way that, on average, for a measuring period of 1 year and 1 kg of target mass, one event which appears outside the band shows up in the tungsten band, too.[7] Due to the relative large overlap of the oxygen band with the tungsten band, it has to be expected, that these events cannot be distinguished. For the interaction rate in the oxygen band only an upper limit is known, see below. In this case, it is assumed that only one event appears in a measuring period of 1 year of 1 kg target mass in the oxygen band. For this reason, the oxygen band width is drawn with a 1 σ energy resolution of the light channel for each energy.

In the pictures the quantization of the energy carriers it not taken into account, i.e. mainly the quantization due to the photon statistics is neglected. From this follows, that the distribution of the bands is a Gaussian distribution.

It can be seen, where in the light-phonon-plane WIMP events, and where background events appear. How these background events influence the CRESST-II sensitivity will be discussed in the next section.

2.3.2 Parameters of the CRESST-II Sensitivity

As can be seen in the middle pictures of figure 2.8, events from other bands can leak into the tungsten band. In the overlapped area both types of particles can show up. Physically this means that the ratio of the two measured signals of both channels cannot be distinguished any more with high confidence on an event by event basis. This is natural given by the finite energy resolutions of both measured channels.

To estimate the significance of this overlapped area of the tungsten band for the Dark Matter search, the event rates of the bands have to be taken into account. Is the event rate in the overlapped area of the background band much higher than the WIMP interaction rate in the same area, the overlapping area cannot be used. The signal is then much smaller than the background, thus this part of the tungsten band is not sensitive enough for WIMP detection.

The event rate of the electron band is much higher than the WIMP interaction rate. It is of the order of 10 events per keV, kg, and day of measurement, whereas the expected WIMP interaction rate, on the other hand, is lower than about 1 event per kg and month over the whole energy range of about 40 keV. Wherever

---

[7]The event rate in the electron band is assumed to be 10/keV kg d. Therefore, the width of the band is chosen as 3.7 σ (99.989 %) energy resolution of the light channel.

2.3 CRESST-II Sensitivity

the electron band overlaps the tungsten band, this area is not sensitive for WIMP detection anymore.
For the oxygen band the situation is different. Here the rate is up to now unknown in the CRESST-II experiment. It is only known that the rate in this band, which is expected to be neutron induced, is lower than 1 event per kg and month over the whole energy range of interest from about 10 keV to about 40 keV. The upper limit of the oxygen and the tungsten band are similar. Therefore the overlapping area can be used for WIMP detection.

Coming back to the electron band which can be seen in the middle pictures of figure 2.8, it overlaps the tungsten band at about 10 keV. Below this energy the low rate of WIMP interactions cannot be observed anymore due to the much higher event rate in the electron band. This energy defines the **WIMP identification threshold**.[8] Below this energy transfer the experiment is not sensitive to WIMPs anymore. It reduces the sensitive area at the low energy range and therefore the number of detectable WIMP interactions. On the other hand, as seen in section 1.3.4, for a WIMP identification a certain interaction statistics is necessary. Therefore, the WIMP identification threshold limits the CRESST-II sensitivity.

Another possibility to increase the number of detected WIMP interactions is to enlarge the **target mass**. In CRESST-II this is realized via the installation of more than one detector module.[9]
The best result of CRESST-II published up to now was measured with two simultaneously operated detector modules. At the moment, data is taken with ten newly installed modules, which corresponds to a total target mass of 3 kg. The maximum number of detector modules that can be installed in the current setup is 33, which corresponds to 10 kg of $CaWO_4$ detector target.

The third way to increase the number of detected WIMP interactions is the extension of the **measurement duration**. In principle there is no limitation given. Due to the ongoing research and development, however, a typical duration measurement of about one year has been chosen in the past. On the one hand, within this period the annual WIMP signal modulation can be measured and, on the other hand, it is a reasonable time for the production of improved detector modules. After a measuring period of about one year an upgrade is usually done, after which the next measuring period starts.[10]

> **The number of identified WIMP interactions can be increased via a lowering of the WIMP identification threshold, an increasing of the target mass, and an extension of the measurement duration.**

---

[8] More precisely one can see that the WIMP identification threshold depends on the detected energy in the light channel. For simplification, this is neglected in most cases.
[9] The other possibility to increase the target mass is to use larger target crystals.
[10] One measuring period is called RUN within the CRESST collaboration.

2.4 Motivation of this Work

WIMP events are expected in the tungsten band. Therefore, these events have to be identified for a positive WIMP signal. The tungsten band overlaps with other background bands, as can be seen in figure 2.8. The overlaps depend on the slopes and the widths of the bands. The WIMP sensitive energy range is the energy range of the tungsten band, where the rate of background events is not much larger than the WIMP interaction rate.

> **The determination and optimization of the bands towards a higher WIMP sensitivity is the goal of this work.**

By this the WIMP sensitivity of the CRESST-II experiment can be increased.

To determine the overlap of the tungsten band with the other bands the **slopes** of the different bands will be determined in chapter 3 with the help of an independent experiment. The parameters which determine the **widths** of the different bands will be determined in chapter 4 and chapter 5. Slopes and widths fix the band positions in the light-phonon-plane and therefore the overlaps with the tungsten band.

It will be seen that the slope of a band is given by law of nature and can therefore only hardly be changed. On the other hand, the width of the bands can be improved. Possibilities for **improvements** will be discussed and tested in chapter 6. Especially at the energy range of the WIMP identification threshold an improvement in the widths of the bands has two effects, which increase the **sensitivity** of the experiment. First, it **reduces the WIMP identification threshold**, so that more identified WIMP events can be expected. Second, the overlap with up to now not limiting bands is reduced. In this way a better background identification is possible.

Both effects can be seen in the two lower pictures of figure 2.8. These pictures differ from the two above in the way that the bands are five times narrower. It can be seen that the overlap of the electron band and the tungsten band is reduced dramatically. The WIMP identification threshold is reduced from about 10 keV to about 2 keV. With the help of formula (1.5) it can be derived that this changes the fraction of detectable WIMP interactions from about **30 %** to about **80 %**. Such a strong effect is due to the fact that the scattering rate for lower transferred energies increases more than exponentially, cp. chapter 1.3.3. This improvement is equal to an increase of the target mass or the measurement duration of more than a factor of 2.5. The second effect can be seen with the oxygen band. A discrimination of oxygen events, which are mainly neutron induced, down to about 10 keV would be achieved.

Chapter 3

Quenching Factors of CaWO$_4$

The previous chapter has shown that WIMP events are expected in the tungsten band. This tungsten band overlaps partially with other bands which are all background originated. For this reason, in the areas of overlap, a reliable WIMP identification is sophisticated due to their very low interaction rate, cp. section 1.3.1. The area of overlap is usually not sensitive enough for the WIMP detection.
Therefore, the parameters which determine these overlap areas are of interest. As seen in the previous chapter, the overlaps are caused by the widths and slopes of the different bands. In this chapter the factor determining the **slopes** of the bands is presented.

In the following section, the slope of a band in the light-phonon-plane will be parameterized by the **quenching factor**. Quenching factors can be measured generally in independent experiments. Recent data of such a **measurement** will be presented. In section 3.2, a **physical explanation** of the quenching factors, and therefore of the origin of the different band slopes in the light-phonon-plane, is presented. Finally, in section 3.3, from this explanation a conclusion about the **linearity of the bands** will be deduced.

3.1 Quenching Factors

3.1.1 Quenching Factor Definition

In section 2.3.1, the so-called **quenching factor** (QF) of a band in the light-phonon-plane has been defined as the ratio of its slope to the slope of the electron/gamma band. The slope of the electron/gamma band is used as reference. Analogously, taking into account a possible energy dependence of the quenching factor, a more general definition is used in this section: The **ratio** of the amount of light produced by an energy deposition E_{dep} of **particle** X in the target crystal and the amount of light produced by an identical energy deposition E_{dep} of **gamma radiation** is defined as quenching factor.

$$\text{QF}_X(E_{dep}) := \frac{\text{light produced by particle } X \text{ of energy } E_{dep}}{\text{light produced by gamma radiation of energy } E_{dep}}$$

Since the light produced by a particle is usually less than the light of gamma events, this ratio is called quenching factor.

As an example, in the first picture of figure 2.8 one can get the quenching factor of alpha-particles from the slope: For an energy deposition of 50 keV of an alpha-particle in the target, on average a 10 keV signal is measured in the light channel. Comparing this with the light produced by electrons of same energy (50 keV) results in a quenching factor for alpha-particles of:

$$\mathrm{QF}_\alpha = \frac{10\,\mathrm{keV}}{50\,\mathrm{keV}} = \frac{1}{5} = 0.2$$

In the CRESST-II experiment, the knowledge of the quenching factors is used for the data analysis: In a dark matter measurement, the position of the gamma band is well known due to the background events. On the other hand, the position of the tungsten band, where WIMP events are expected to appear, is not visible, since only few events appear in this band, cp. section 1.3.1. Therefore, the position of this band can be determined from the position of the gamma band and the knowledge of the tungsten quenching factor, which gives the relative positions of these two bands. In this way the mean value of the tungsten band can be fixed. The width of the tungsten band can be derived from the measured gamma band width, too. This procedure fixes the area in the light-phonon-plane where WIMP events are expected with the help of the **tungsten quenching factor**.

3.1.2 Quenching Factor Measurement for CaWO$_4$

The necessity of the quenching factor knowledge for CaWO$_4$, of at least the element tungsten, for the CRESST-II experiment has been motivated. In this section, recent results of a quenching factor measurement will be presented.

Section 2.3.1 has shown that for each particle interacting electro-magnetically a band shows up in the light-phonon-plane. These bands, and in the same way their relative position to the gamma band, i.e. the quenching factors, can be measured in two different ways:

One possibility is that an **internal** component of the CaWO$_4$ target crystal loses energy. Therefore an energy E_{dep} has to be transferred to one of the CaWO$_4$ components. This is usually realized by neutrons, since they are able to fly through the target crystal by interacting only once. By scattering off, for example, an oxygen nucleus of the crystal, a neutron can transfer a part of its energy onto this nucleus. By measuring the phonon channel or the neutron kinematics, the deposited energy can be determined. The light channel measurement, together with a gamma reference measurement of same energy, defines the quenching factor at this energy. With the help of this measurement principle, it is possible to determine the quenching factors of **all CaWO$_4$ components**: e$^-$, O, Ca, and W.

3.1 Quenching Factors

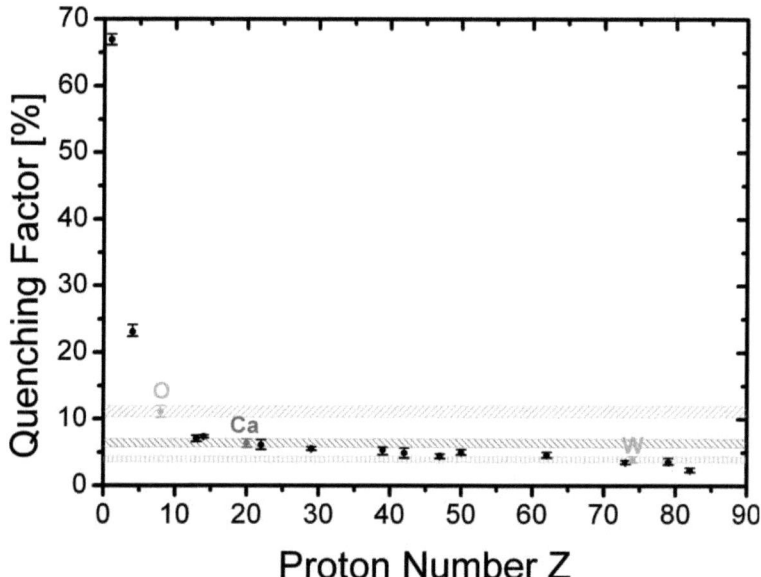

Figure 3.1: In this plot, CaWO$_4$ quenching factors for 18 keV are shown for different elements depending on their proton number Z. For small proton numbers, as for light elements as oxygen, it decreases rapidly, whereas for large proton numbers, i.e. heavy elements as tungsten, the quenching factors approach to a constant value. The three values important for CRESST-II (O, Ca, and W) are marked.

A second possibility is the total energy deposition of an **external** particle which interacts electro-magnetically, as it is the case, for example, for alpha particles. The energy of the absorbed particle, E_{dep}, can be determined by measuring the phonon channel or from the knowledge about the incident particle. The light channel measurement enables the quenching factor determination. In this way, it is in principle possible to derive the quenching factors for CaWO$_4$ of **all electro-magnetically interacting particles**, i.e. for all bands which can show up in the light-phonon-plane.[1]

Recent measurement results of quenching factors for CaWO$_4$ for an energy of 18 keV at room temperature[2] (300 K) can be seen in figure 3.1 and table 3.1,

---

[1] It should be noted that not electro-magnetic interacting particles, as neutrons and WIMPs, mainly contribute to recoils of one component. These are, in case of these two particle types, oxygen and tungsten nuclear recoils respectively, cp. section 2.3.1.

[2] It is assumed that quenching factors of CaWO$_4$ do not have a strong temperature dependence in the energy range relevant for CRESST-II. This is due to the fact that the quenching factors are defined as ratio of two light signals. Although one light signal is temperature dependent, for the ratio of two of these signals the temperature dependence should cancel. This could be confirmed in [30].

Element	Proton Number Z	Quenching Factor [%]
H	1	$100 \cdot 1/1.494^{+0.018}_{-0.018} = 66.93^{+0.819}_{-0.811}$
Be	4	$100 \cdot 1/4.335^{+0.137}_{-0.187} = 23.07^{+1.038}_{-0.707}$
O	8	$100 \cdot 1/9.020^{+0.805}_{-0.684} = 11.09^{+0.909}_{-0.908}$
Al	13	$100 \cdot 1/14.184^{+1.003}_{-0.903} = 7.05^{+0.479}_{-0.466}$
Si	14	$100 \cdot 1/13.605^{+0.415}_{-0.407} = 7.35^{+0.227}_{-0.218}$
Ca	20	$100 \cdot 1/15.675^{+1.789}_{-1.386} = 6.38^{+0.619}_{-0.653}$
Ti	22	$100 \cdot 1/16.479^{+2.070}_{-1.900} = 6.07^{+0.791}_{-0.677}$
Cu	29	$100 \cdot 1/18.018^{+0.876}_{-0.824} = 5.55^{+0.266}_{-0.257}$
Y	39	$100 \cdot 1/18.695^{+2.866}_{-1.338} = 5.35^{+0.412}_{-0.711}$
Mo	42	$100 \cdot 1/20.463^{+3.655}_{-2.671} = 4.89^{+0.734}_{-0.741}$
Ag	47	$100 \cdot 1/22.366^{+2.064}_{-1.411} = 4.47^{+0.301}_{-0.378}$
Sn	50	$100 \cdot 1/20.139^{+1.529}_{-1.762} = 4.97^{+0.476}_{-0.350}$
Sm	62	$100 \cdot 1/21.887^{+1.835}_{-1.995} = 4.57^{+0.458}_{-0.353}$
Ta	73	$100 \cdot 1/28.302^{+1.915}_{-2.167} = 3.53^{+0.293}_{-0.224}$
W	74	$100 \cdot 1/25.566^{+3.154}_{-2.784} = 3.91^{+0.478}_{-0.430}$
Au	79	$100 \cdot 1/28.135^{+3.084}_{-4.019} = 3.55^{+0.592}_{-0.351}$
Pb	82	$100 \cdot 1/42.579^{+6.482}_{-5.971} = 2.35^{+0.383}_{-0.310}$

Table 3.1: Quenching factor measurement results of $CaWO_4$ for 18 keV at room temperature [29]. These values are plotted in figure 3.1.

data taken from [29]. The technique used in this measurement produces external particles with an energy of $E = 18\,\text{keV} = E_{dep}$. These particles are produced by shooting with a laser onto a material sample, as for example tungsten. Single charged ions are produced and accelerated via a potential difference of 18 keV. The ions are deflected into a $CaWO_4$ crystal. In this way these particles, for example tungsten ions, deposit their energy in the target crystal. The light output of the target crystal is measured with a photomultiplier. The comparison of this light output with the one of gammas of 18 keV results in the shown quenching factor. This is done for several elements from hydrogen to lead. For further description of the measurement principle see [30] or [31]. In this context the focus will be on the results and conclusions which can be derived from this measurement.

It can be seen that for an increase of the proton number the quenching factor decreases monotonically, rapidly for light elements and smoothly for heavy elements. The three elements important for CRESST-II (O, Ca, and W) are marked. It can be seen that they have **significantly different quenching factors** which, in principle, enables the CRESST-II experiment to distinguish them. For example, if an energy of 18 keV is deposited in the target crystal, in average a light signal (LS) of $QF_X \times 18\,\text{keV}$ will be measured. X is the component of $CaWO_4$ onto

3.2 Quenching Factor Explanation

which the energy is transferred:

$$QF_e = 100.00\,\% : \quad LS = 18.00\,\text{keV}$$
$$QF_O = 11.08\,\% : \quad LS = 1.99\,\text{keV}$$
$$QF_{Ca} = 6.38\,\% : \quad LS = 1.15\,\text{keV}$$
$$QF_W = 3.91\,\% : \quad LS = 0.70\,\text{keV}$$

If a measured light signal of an event can be assigned to or excluded from a quenching factor, considering energy resolution and quenching factor uncertainty, information about the component onto which the energy was transferred, can be obtained. From this knowledge conclusions can be drawn about the incident particle which can be identified as background or considered for a WIMP signature identification, cp. chapter 2.

In the case above, for example, a signal in the phonon channel of 17.8 keV and a signal in the light channel of $LS = 20.5\,\text{keV}$ can be measured. This results in a light to phonon ratio of 1.15, which can be related to the electron quenching factor of one. For this reason it can be concluded that the energy was distributed in the crystal by an electron. Therefore the incident particle was gamma or beta radiation, in any case a background particle and not a WIMP. This event is identified as background event.

> **The quenching factor measurement shows that it is in principle possible to identify the component of $CaWO_4$ onto which energy is transferred and thus to get information about the incident particle.**

3.2 Quenching Factor Explanation

3.2.1 Saturation of the Light Production

The quenching factor measurement shows that the **detected** amount of light depends on the proton number Z of the component which distributes the energy in the target crystal, see figure 3.1. This implies that the fraction of deposited energy which is **transformed** into scintillation light depends on the particle. The reason for this behavior will be discussed in this section.

In figure 3.2, the averaged number of detected photons is shown, as a function of the proton number Z. This is the same plot as figure 3.1 of the quenching factors; the difference is that these values are multiplied by the average number of detected photons of an 18 keV gamma event. Additionally, the averaged simulated path lengths of these nuclei in $CaWO_4$ with a starting energy of 18 keV are plotted [32]. A clear **correlation** between the detected photon number and the simulated path length can be seen.

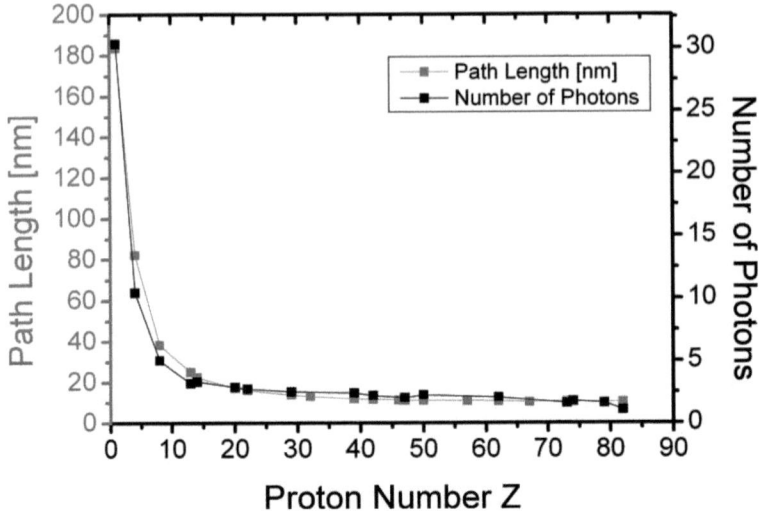

Figure 3.2: In this figure the simulated path length of different elements with a starting energy of 18 keV in CaWO$_4$ can be seen. On the right hand side, the number of detected photons produced by these nuclei of same energy are plotted. A clear correlation of the two values can be seen. The number of detected photons produced by a nucleus depends strongly on the path length of this nucleus.

|| **The amount of scintillation light produced in CaWO$_4$ depends strongly on the path length of the nuclei.** ||

To underline this, in figure 3.3 the ratios of these two values are shown. These ratios are nearly independent of the proton number Z.

On the other hand, all simulated nuclei in CaWO$_4$ have an incident energy of 18 keV, but their average path length differs by more than one order of magnitude, see figure 3.2. From this can be concluded that the energy loss per path length (dE/dx) differs strongly between the nuclei, depending on the number of protons. This can by seen in figure 3.4, too. In this figure dE/dx is shown for the three nuclei oxygen, calcium, and tungsten for energies below 200 keV. It can be seen that it differs by more than one order of magnitude. dE/dx is larger for nuclei with higher proton number; therefore their total path length is shorter.

The correlation between path length and photon number can be explained by a **saturation** effect: Although dE/dx differs between the nuclei by more than one order of magnitude, cp. figure 3.4, the number of detected photons per path length (dL/dx) is constant, cp. figure 3.3. One can conclude for **nuclei** that dL/dx is independent of dE/dx.

3.2 Quenching Factor Explanation

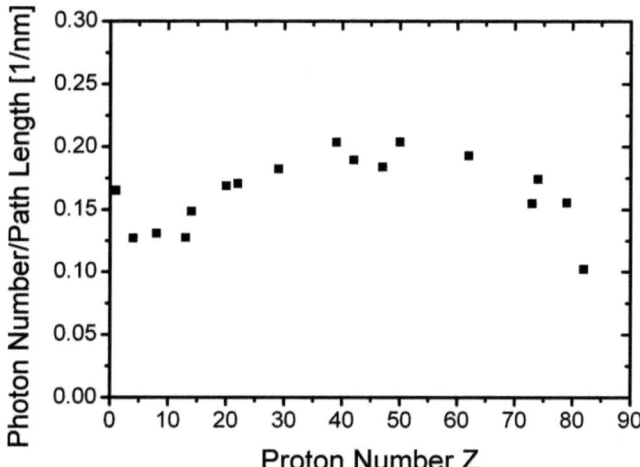

Figure 3.3: In this proton number dependent plot the ratio of the detected number of photons to the simulated path length of different nuclei is shown. This is the ratio of the two values plotted in figure 3.2. The number of photons per path length is approximately independent of the proton number Z.

> Independent from dE/dx, for nuclei always the same amount of energy per path length is transformed into scintillation light.

This saturation concept is consistent with the following consideration: One can see in figure 3.3 that about 0.1 - 0.2 photons per nm path length of a nucleus in $CaWO_4$ are detected. Assuming a quantum efficiency of 20 %, this results in a light production of 0.5 - 1 photons per nm path length. On the other hand, the average distance between two WO_4^{2-}-complexes, which are assumed to produce the scintillation light, is about 0.5 - 1 nm, depending on the direction. For this reason the measurement is consistent with the picture that nuclei in $CaWO_4$ excite each WO_4^{2-}-complex on their path length, where each of the complexes produces one scintillation photon.

3.2.2 Linearity of the Light Production

Compared to nuclei, the situation for events caused by gammas or electrons is totally different. As can be seen from figure 3.4, dE/dx of electrons in $CaWO_4$ is about two orders of magnitude smaller than for nuclei. This is reflected in a much longer path length [33]. For electrons of less than about 20 keV dE/dx increases towards lower energies.

This energy dependent dE/dx (E) of electrons [33] and the energy dependent

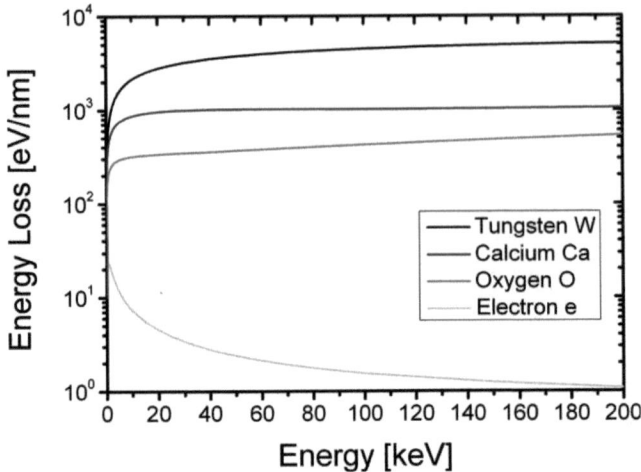

Figure 3.4: In this plot dE/dx(E) of tungsten (W), calcium (Ca), oxygen (O), and electrons (e) in CaWO$_4$ are plotted [32][33]. For the nuclei, dE/dx is much larger than for electrons. dE/dx is roughly constant for energies above 10 - 20 keV, i.e. independent of the particle's energy. On the other hand, it strongly depends on the number of charge carrier (proton, electron) of the particle.

dL/dE (E) of electrons in CRESST-II[3] [34] can be combined to plot 3.5. Therefor the energy dependent light L(E), taken from figure 3.8, is differentiated with respect to the energy:

$$\frac{dL}{dE}(E)$$

For each data point one point in plot 3.5 is determined:

$$(x;y) = \left(\frac{dE}{dx}(E); \frac{dL}{dE}(E) \cdot \frac{dE}{dx}(E)\right) = \left(\frac{dE}{dx}; \frac{dL}{dx}\right)$$

For this reason, the plotted points reflect the detected light of electrons per simulated path length (dL/dx) as a function of dE/dx. A **linear** dependence can be seen, i.e. the more energy is deposited in a small volume of the target crystal, the more energy is transformed into scintillation light in this volume. The light output is linear with respect to the deposited energy.

> For electrons the fraction of deposited energy which is transformed into scintillation light is constant, independent of dE/dx.

---

[3] In figure 3.8 a similar picture can be seen. In this picture the energy dependent ratio of the light to energy is plotted.

3.2 Quenching Factor Explanation

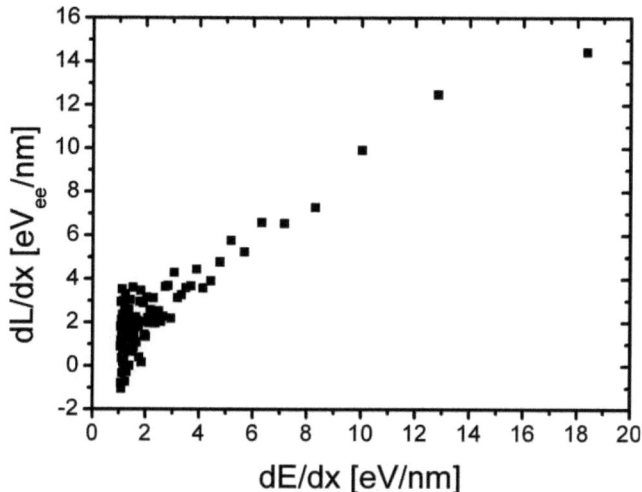

Figure 3.5: In this plot dL/dx is shown for electrons depending on dE/dx. Both values show a linear correlation, i.e. the more energy is deposited in a small volume the more light is produced in this volume. Therefore, in case of electrons, the fraction of deposited energy transformed into scintillation light is constant.

3.2.3 Energy Deposition and Light Production

In this section the two previous observations are combined. Section 3.2.1 has shown that for **nuclei** the **amount** of energy deposited in a small volume which is transformed into scintillation light is **constant**, independent of dE/dx. In contrast to section 3.2.2, which has shown that for **electrons** the **fraction** of energy transformed into scintillation light is **constant**, independent of dE/dx.

Figure 3.6 contains all derived values in one plot. The electron values from figure 3.5 and the values from nuclei derived from figure 3.3.

For the nuclei it is assumed that the 18 keV energy is deposited homogeneously all over the simulated path length. I.e. it is neglected that a fraction of the energy is distributed by secondary nuclei and that dE/dx is energy dependent. To determine dL/dx of the nuclei, the results from figure 3.3 cannot be taken directly, since the fraction of detected light differs significantly from the light measurement of the electrons. In the case of nuclei the target is not optically coupled to the light detector by vacuum grease and the target is at room temperature. In the case of the electrons, the light detector is weakly coupled to the target and the measurement is done at ultra-low temperatures. Additionally, both measurements are done with different light detectors with different quantum efficiencies. A direct comparison of both absolute values is not possible.

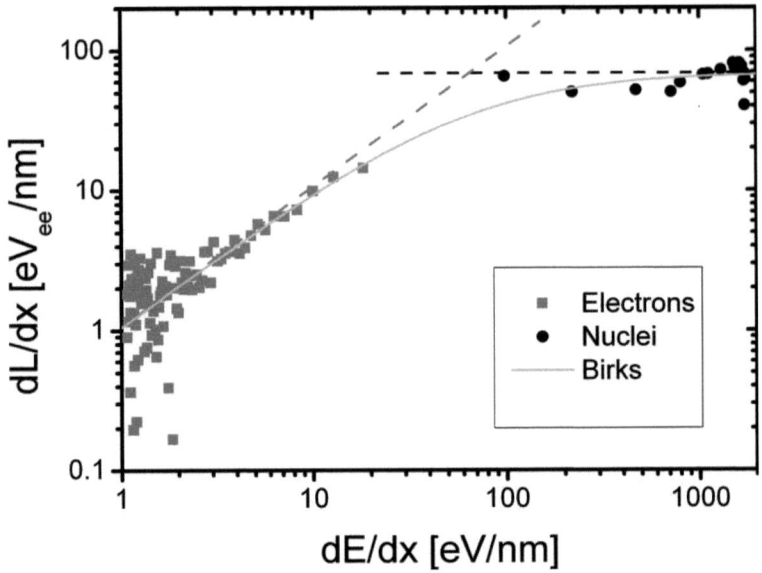

Figure 3.6: In this figure derived values of dL/dx depending on dE/dx are shown. The electron events are located at small dE/dx, where the light production is proportional to the energy deposition. The nuclei are placed at large dE/dx, where the light production is saturated. All these events can be described by Birk's formula. This formula fit is plotted, too. The two limits for small and large dE/dx are shown as straight dashed lines.

For this reason, for the nuclei dL/dx is determined with the help of the measured quenching factors. dE/dx is multiplied by the quenching factor. In this way dL/dx can be derived. These values are plotted in figure 3.6.

In this figure can be seen, that **electrons** are in the range where dL/dx is proportional to dE/dx:

$$\frac{dL}{dx} \sim \frac{dE}{dx}$$

For **nuclei** dL/dx is, on the contrary, independent of dE/dx. The light production is saturated:

$$\frac{dL}{dx} = \text{const.}$$

This behavior can be expected naturally: Imagine a small volume of a scintillator in which energy is deposited. For small energy depositions the amount of produced light will increase for rising energy depositions. If all light producing mechanisms are activated, an increase of the energy deposition will not increase the amount of produced light anymore.

3.2 Quenching Factor Explanation

For organic scintillators such a behavior is well known and often described by Birk's formula:

$$\boxed{\frac{dL}{dx} = \frac{A \cdot \frac{dE}{dx}}{1 + B \cdot \frac{dE}{dx}}}$$

This is the simplest possible description of the above behavior within one formula. Fitting this formula to the derived values gives the two free parameters A and B:

$$A = 1.065$$
$$1/B = 65 \, \text{eV/nm}$$

Parameter A describes the proportionality of the light production to the energy deposition for small dE/dx, as it is the case for electrons:

$$\frac{dL}{dx} \stackrel{\frac{dE}{dx} \ll \frac{1}{B}}{\approx} A \cdot \frac{dE}{dx}$$
$$\Rightarrow L \approx A \cdot E$$

The detected light L is proportional to the total deposited energy E_{dep}. Parameter A depends not only on physical properties, as the fraction of deposited energy which is transformed into scintillation light, it also depends on technical properties such as the coupling and properties of the light detector.
A/B reflects the maximum dL/dx for large dE/dx, as it is the case for nuclei:

$$\frac{dL}{dx} \stackrel{\frac{dE}{dx} \gg \frac{1}{B}}{\approx} \frac{A}{B}$$
$$\Rightarrow L \approx \frac{A}{B} \cdot x$$

The detected light is proportional to the path length of the particle in the scintillating crystal.

Both limits are shown as straight dashed lines in figure 3.6. The intersection point of these two lines is given at:

$$\frac{dE}{dx} = \frac{1}{B}$$

Parameter B is called **Birk's parameter**. It describes the value of dE/dx for which half of the maximum dL/dx is produced:

$$\frac{dL}{dx} \stackrel{\frac{dE}{dx} = \frac{1}{B}}{\approx} \frac{1}{2}\frac{A}{B}$$

Assuming that the relation between **dL/dx** and **dE/dx** is well known, cp. figure 3.6, and is independent of the type of particle (electrons and nuclei), then this relation and the knowledge of the **dE/dx (E)** of a particle in a scintillator, cp. figure 3.4, fix the relation between total detected light and total deposited energy: **L(E)**. This relation describes the absolute band position of this particle in the light-phonon-plane.

> The relation between dL/dx and dE/dx of a material
> fixes the positions of the bands in the light-phonon-plane.

Assuming Birks formula describes well the scintillator behavior,[4] all band positions in the light-phonon-plane are fixed by the so-called **Birks factor** B and the knowledge of dE/dx (E) of the particles in a scintillator, cp. figure 3.4.

However, it turns out that Birks formula does not describe the band positions in the light-phonon-plane well. Reasons for this can be, on the one hand, that the path length of the nuclei is underestimated due to the neglected secondary recoils. On the other hand, that Birk's formula is too imprecise for the description of the relation between the energy loss and light production in $CaWO_4$.

3.3 Energy Dependence of the Bands

Energy Dependence of the Nuclei Bands

The energy dependence of nuclei bands in the light-phonon-plane can be estimated simply: As seen in section 3.2.1, for **nuclei** dL/dx is independent of dE/dx. For this reason, the absolute amount of light produced by a nucleus L is proportional to the path length x:

$$L \sim x$$

The path length x of different nuclei and energies in $CaWO_4$ can be simulated [32]. The results for beryllium (Be), oxygen (O), and tungsten (W) for energies below 50 keV can be seen in figure 3.7. The absolute path length of the simulated nuclei is proportional to the energy of the incident particle:

$$x \sim E$$

For this reason, it can be expected that the light produced by nuclei in $CaWO_4$ is proportional to their initial energy:

$$L \sim E$$

> The nuclei bands in the light-phonon-plane
> are expected to be lines through the origin.

---

[4]This means, among other things, that the formula holds in the same way independent of the kind of stopping power: Whereas electrons transfer their energy onto other electrons (electron stopping), nuclei transfer their energy mainly onto other nuclei (nuclear stopping) for CRESST-II relevant energies. It is also neglected that secondary nuclei and electron recoils produce light, whereas their path length in the target is not taken into account.

3.3 Energy Dependence of the Bands

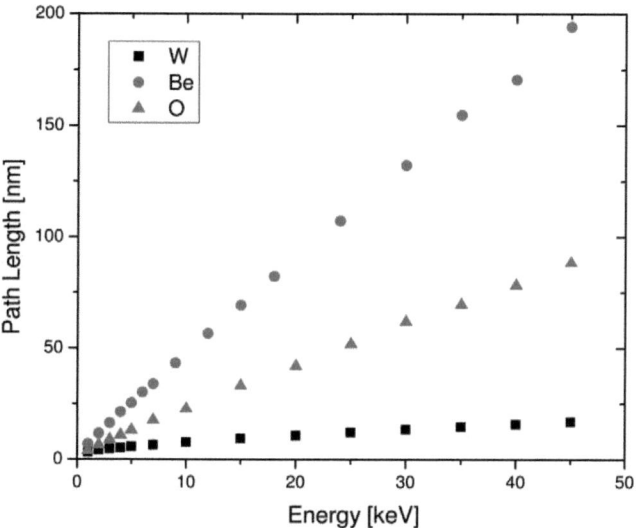

Figure 3.7: In this figure the path lengths of three different elements in CaWO$_4$ with different initial energies (< 50 keV) are simulated [32]. The path lengths are to first order, proportional to the initial energy, i.e. dE/dx of these nuclei is not strongly dependent on their energy.

Energy Dependence of the Electron Band

For electrons the situation is different, as seen in section 3.2.2. In CaWO$_4$ the fraction of deposited energy which it transformed into scintillation light is, for electrons, independent of dE/dx. If this fraction is, for example, 2 %, independently of dE/dx always 2 % of the deposited energy will be transformed into light. The produced light is proportional to the deposited energy:

$$L \sim E$$

> The electron band in the light-phonon-plane
> is expected to be a line through the origin.

This linearity can be tested by a measurement of the energy dependence of the electron band. Therefore data taken at Gran Sasso have been analyzed [34]. The fraction of the deposited energy which is transformed into light (light to phonon ratio) is averaged in 2 keV bins. This energy dependent ratio can be seen in figure 3.8. Additionally, a dashed line shows the expected energy dependence of a constant slope of one for electrons. For large energies the fraction of light increases, whereas it decreases for small energies. For zero energy the ratio

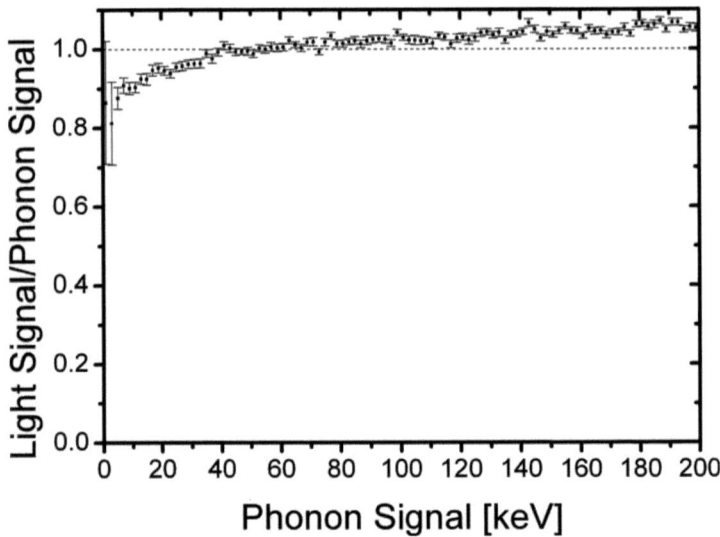

Figure 3.8: In this figure the ratio of the light signal to the phonon signal produced by gammas/electrons in CaWO$_4$ is shown. For different energies this ratio is averaged in 2 keV bins. A constant slope of one, as the light channel is calibrated for 122 keV gamma events, of the electron band in the light-phonon-plane is marked as dashed line. The slight discrepancy from this line indicates the energy dependence of the electron band, which can be explained by Birk's formula and dE/dx (E) of electrons.

is about 0.85. This deviation from the dashed line is called **scintillator non-proportionality**.

It can be explained from the two figures 3.4 and 3.6. In figure 3.4 can be seen that dE/dx of electrons increases for low energies. Combining this information with figure 3.6 shows that electrons of low energies leave the linear range of light production. The fraction of energy transformed into light is reduced. This is reflected in figure 3.8 at low energies.

Chapter 4
Energy Resolution of the Light Channel

As seen in chapter 2, the tungsten band in the light-phonon-plane, where WIMP events are expected, overlaps partially with background bands, see figure 2.8. The overlap depends on the widths and different slopes of the bands. This chapter discusses the **widths** of the bands.

The widths appear naturally since, although exactly the same energy E_{dep} is deposited in the detector target, different energies E can be measured. Reasons for this can be, for example, a difference in the energy transport or in the noise influence. Therefore, each channel has a finite energy resolution which is usually energy dependent. The measured **energy resolution** can generally be defined as **full width at half maximum (FWHM)** δE per **mean measured energy** $\langle E \rangle$:

$$\boxed{Energy\ Resolution := \frac{\delta E}{\langle E \rangle}}$$

Here, the FWHM is only <u>one</u> possible choice to specify the measurement uncertainty of an energy. The **energy resolutions** of the phonon and the light channel describe the **widths** of the bands in the light-phonon-plane.

Out of the two measured channels, the light channel has the worse energy resolution. This can be seen, for example, in a calibration measurement. A typical calibration measurement is shown in figure 4.1. A $CaWO_4$ target was irradiated with 122 keV and 136 keV gammas. Both channels detected coincident signals. In the left picture of this figure these signals are plotted in the light-phonon-plane. The discrete lines at the two calibration energies can be seen as longish areas in the plane. Additionally, at lower energies two further lines appear which are due to escape events. I.e. the gamma quantum is absorbed via photoelectric effect, the electron vacancy is filled and the thereby emitted X-ray quantum escaped from the target crystal. This way, the energy of the X-ray is lost (about 60 keV) and a reduced energy is detected.

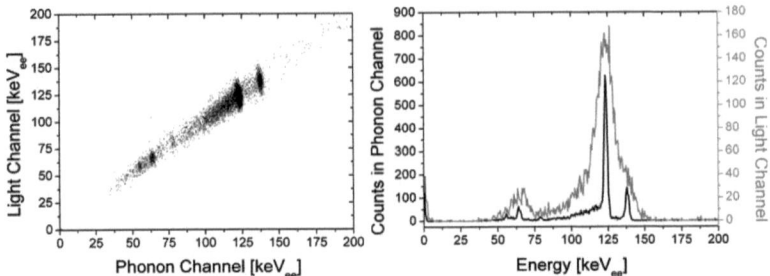

Figure 4.1: A CaWO$_4$ target crystal was irradiated by 122 keV and 136 keV gammas. Both pictures show the signals measured by the two channels. In the left hand picture, the coincidently measured signals are plotted in the light-phonon-plane. In the right hand picture, both channel's histograms are shown. This plot shows nicely that the energy resolution in the light channel is much worse than the one of the phonon channel. Therefore, the width of the band, see the plot of the light-phonon-plane on the left hand side, is dominated by the energy resolution of the light channel.

If all different energies had been measured, one would observe a constant band instead of these discrete longish lines. On the other hand, from this longish line one can see that the width of the continuous band is dominated by the energy resolution of the light channel. The difference in the energy resolution of the two measured channels can also be seen on the right hand side. There histograms of both measured channels are shown. The difference of the energy resolutions of the two measured channels can be seen directly. Whereas the phonon channel clearly resolves the two gamma energies of 122 keV and 136 keV, this is not the case for the light channel.

The main reason for the different energy resolutions of the two channels is that only a few percent of the deposited energy is converted into scintillation light.[1] The main part of the deposited energy is transformed into phonons [32]. For this reason the light signal detection is more challenging and the light channel ends up with a worse energy resolution.

The previous calibration data belong to the electron/gamma band; for the oxygen, the calcium, the tungsten, and the phonon only band an even smaller fraction of the deposited energy is transformed into scintillation light, cp. chapter 3. Thus the influence of the relative light channel resolution onto their band widths is even larger. From this one can conclude:

> **The energy resolution of the light channel dominates the widths of the bands of interest.**

---

[1] Typically in the light channel of CRESST-II between 1% and 2.5% of the energy which is deposited by a gamma in the target is detected as scintillation light [35].

This fact can be seen on the basis of the two lower pictures of figure 2.8. In contrast to the two middle pictures with typical energy resolutions, the energy resolution of the light channel is improved by a factor of five for the two lower pictures, resulting in much narrower bands. This way the overlap of different background bands with the tungsten band can be reduced substantially. On the one hand, the overlap of the limiting electron band is pushed to lower energies, on the other hand, oxygen recoils can be distinguished down to about 10 keV. With such an improvement the WIMP sensitivity of the experiment could be increased significantly.

With the help of the calibration measurement shown above, the influence of the energy resolution of the light channel onto the band widths can be illustrated at energies above 100 keV. For the CRESST-II sensitivity, on the other hand, the WIMP identification threshold is important. This energy threshold is defined by the overlap of the tungsten band and the electron band. Therefore the widths of these two bands **in the energy range of the WIMP identification threshold is of interest for the CRESST-II sensitivity**. This will be the focus of this chapter.

In the next section the setup of the light channel is presented in more detail. Afterwards, in section 4.2, the **mean measured energy $\langle E \rangle$** and, in section 4.3, the **measurement uncertainty δE** will be discussed. Thus the parameters, which define the energy resolution $\delta E/\langle E \rangle$ of the light channel are determined. With the help of this information different possible improvements for the CRESST-II sensitivity will be finally discussed and tested in chapter 6.

4.1 Setup of the Light Channel

As described in section 2.2.2, the light channel consists of a **target crystal**, a **reflecting housing**, a **light absorber**, and a **thermometer**. In the following these components will be discussed to be able to derive a signal development in the light channel.

4.1.1 Target Crystal

The detector target of CRESST-II is a cylindrical crystal with a diameter and height of 40 mm. This shape is chosen to simplify the crystal production. $CaWO_4$ is used as target material. This material was chosen since it has an element with a high atomic mass number ($A_W = 184$ u) which is expected to have a high WIMP-interaction rate, cp. section 1.3.1, a relatively low intrinsic radioactivity, combined with a relatively high light output at ultra-low temperatures ($T < 100$ mK). In figure 4.2, a measured scintillation light **emission spectrum** of a $CaWO_4$ crystal can be seen [36]. This measurement was done at 77 K; it is not expected that the spectrum shifts significantly going on to lower temperatures, since it nearly does

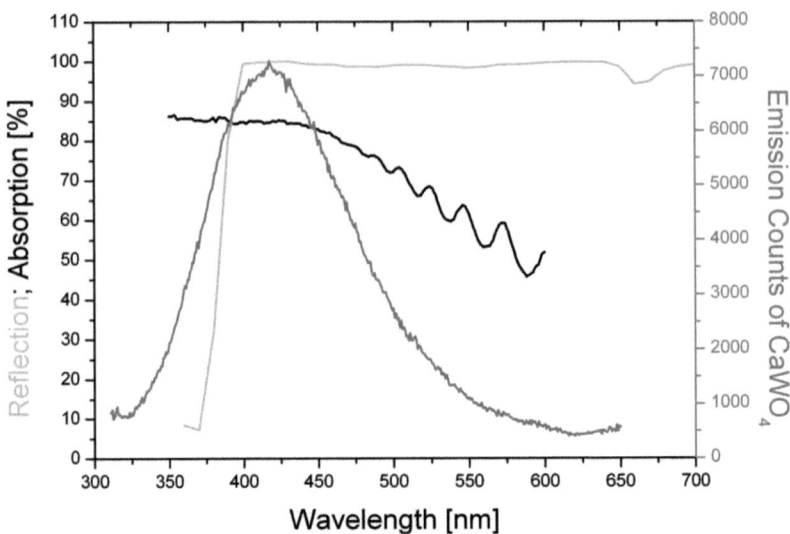

Figure 4.2: In this figure three graphs depending on the light wavelength are shown. The emission spectrum of a CaWO$_4$ crystal can be seen in gray [36]. The light gray curve represents the reflection properties of the light reflecting foil surrounding the detectors [37]. Below about 400 nm photons are mainly absorbed by the foil. These photons can be emitted again with reduced energy due to the scintillation properties of the foil. The absorption probability of the light absorber is plotted in black [38].

not change between 300 K and 77 K [36]. The main emission is in the range of 370 nm to 480 nm with a maximum at about 420 nm. For this wavelength range the reflection properties of the reflecting housing and the absorption properties of the light absorber should be as high as possible to transport a huge amount of the emitted scintillation light from the crystal to the light detector.

Due to the high index of refraction of CaWO$_4$, n$_{CaWO_4}$ = 1.92, **total internal reflection** appears already at an incident angle of $\geq 31°$. The relatively high internal reflection probability combined with the high geometric symmetry of the crystal, which causes recurrent incident angles in subsequent reflections, leads to trapped light in the target crystal. For this reason **roughening** of one front side of the crystal was introduced [36]. The roughness is of the order of 10 µm. Thus the incident angle changes for the one front side. Hence the transmission probability and the reflection angles are changed. The total amount of emitted light is enlarged significantly. The effect of the roughening of the front side of a CaWO$_4$ crystal can be seen on the right hand side of figure 2.6. In contrast to that the surface on the left hand side is polished.

4.1.2 Reflecting Housing

The reflecting housing of a detector module consists mainly of a **highly light reflecting polymeric foil**. It is mounted at a distance of about 5 mm to the crystal and light detector. A larger distance was not realized due to the limited experimental volume. The foil is named VM2002 and its reflecting properties were measured under an angle of incidence of 10° at 300 K. The measurement's results can be seen in figure 4.2 [37]. Due to the uncertainty of the calibration measurement, values slightly above 100 % have been detected. However, the reflection probability for wavelengths above 400 nm is close to 100 % which is confirmed by the producer. Below 400 nm the absorption probability of the foil increases rapidly. These absorbed photons, on the other hand, can be re-emitted in form of scintillation light with reduced energy.

4.1.3 Light Absorber

The standard CRESST-II light absorber is a 460 µm thick sapphire crystal with a diameter of 40 mm. One face is covered with a 1 µm epitaxially grown silicon layer, which serves as light absorber. By default the sapphire side of this absorber is facing the target crystal due to the higher **absorption probability** for incoming light. The wavelength dependent absorption probability can be seen in figure 4.2 [38]. The interference pattern at larger wavelengths is caused by the 1 µm thick silicon layer.

4.1.4 Thermometer

The last component of the light channel setup is a thermometer which measures the temperature rise of the light absorber. As in the case of a phonon channel, a transition edge sensor (TES) introduced in section 2.2.1 is used. A picture of the thermometer structure which is used for the light channel, can be seen in figure 4.3(a). At the top center of the picture a **tungsten thermometer** with a size of 450 µm × 300 µm and 200 nm thickness can be seen. On each of the short sides aluminum films with a size of 1 000 µm × 500 µm and 1 000 nm thickness are placed. On the lower end an additional cooling- (gold) and heating-structure (aluminum) for the thermal stabilization of the thermometer in the transition is placed. The cooling structure is also used to transport detected energy out of absorber and thermometer. The blackish lines, which can be seen, are electrical and thermal connection bond wires.

To read out the thermometer resistance change, a circuit including a **SQUID** (superconducting **qu**antum **i**nterference **d**evice) is used, which can be seen in figure 4.3(b). The circuit consists of the tungsten thermometer in parallel to a SQUID input coil and a shunt resistance. The total bias current I_{Tot} through the two parallel branches of the setup is constant. Thus the bias current reduction through the thermometer due to a resistance increase is equal to the bias current increase through the constant shunt resistor, which is detected via the SQUID.

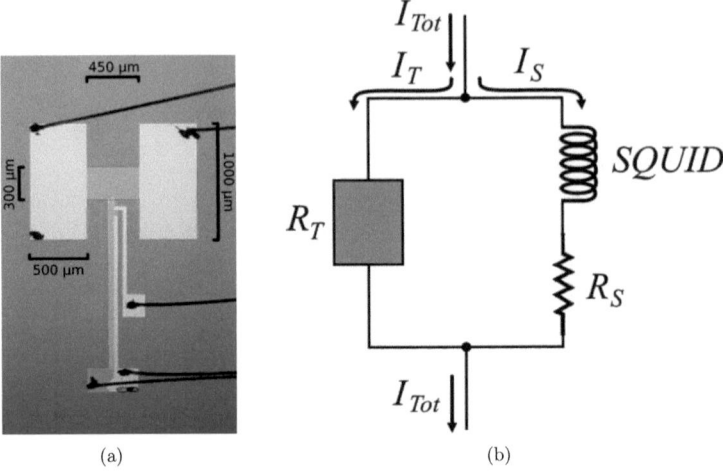

Figure 4.3: On the left hand side the thermometer structure of a light detector of CRESST-II can be seen. The temperature sensitive thin tungsten film is located at the top center. On its right and left aluminum films are connected, which should enlarge the collected energy in the thermometer, cp. section 5.3.4. On the lower side of the thermometer a gold cooling and an aluminum heating structure is placed to stabilize the thermometer thermally. On the right hand side a schematic drawing of the read out circuit is shown. A constant bias current is split into the thermometer and the SQUID branch. This splitting depends on the thermometer resistance R_T and is measured by the SQUID.

4.2 Mean Measured Energy

On the basis of previously described setup the signal development in the light channel can be summarized as follows: After an **energy deposition** in the detector target, not more than a few percent of this energy are transformed into scintillation light. Part of these **photons** can escape the crystal and end up in the light absorber, where this energy is transformed into **phonons**. These phonons distribute all over the absorber and the **thermometer**. The temperature rise of the thermometer is measured by a **SQUID** based electronics.

As seen, the average measured energy[2] E in the light channel following an energy deposition E_{dep} in the target crystal depends on many parameters. These parameters will be summarized first and after that discussed in the following sections in detail.

---

[2] For a clear arrangement in the following expectation values are **not** marked by brackets: $E \equiv \langle E \rangle$.

4.2 Mean Measured Energy

- The average fraction of the deposited energy, which is transformed in the target into scintillation light is p (chapter 4.2.1). This fraction depends on the number of produced scintillation photons and their average energy. The absolute energy transformed into scintillation light is then given as:

$$p\, E_{dep}$$

- The average fraction of scintillation light which escapes the crystal and is absorbed by the light detector is q (chapter 4.2.2). This depends on the self-absorption and emission properties of the crystal, the reflectivity of the surrounding housing, the geometric configuration of the crystal, the light detector and the surrounding housing, and the properties of the light absorber.

$$E_{LD} := p\, q\, E_{dep}$$

- The average part of the energy absorbed by the light detector which is transferred to the read out thermometer film and changes the measured temperature is r (chapter 4.2.3). This depends on the phonon evolution, propagation, transmission, and absorption in the light detector itself.

$$r\, E_{LD} := p\, q\, r\, E_{dep}$$

- The average transferred energy into the thermometer, $r \cdot E_{LD}$, heats up the thermometer by ΔT_T^{dep}, which is inversely proportional to the thermometer heat capacity C_T (chapter 4.2.4).[3] For small energy depositions the temperature dependency of the heat capacity is negligible:

$$\Delta T_T^{dep} = \frac{r\, E_{LD}}{C_T} \quad (4.1)$$

- The change of the temperature leads to a change in the resistance of the thermometer ΔR_T dependent on the transition slope m (chapter 4.2.5). For small energy depositions the transition slope is approximately temperature independent:

$$\Delta R_T = m \cdot \Delta T_T^{dep} \quad (4.2)$$

- The change of resistance leads to a change of the bias current through the SQUID, ΔI_S, dependent on the bias current through the thermometer film I_T (chapter 4.2.6), the shunt resistance R_S (chapter 4.2.7) and the thermometer film resistance R_T (chapter 4.2.8). The absolute value of R_T is approximately constant for small energy depositions while a pulse, therefore can be written:

$$\Delta I_S = \frac{I_T}{R_S + R_T} \cdot \Delta R_T$$

---

[3]The total thermometer temperature change ΔT_T is not only affected by the energy deposition; also the change of the bias current heating (electro-thermal feedback) affects it. Hence the thermometer temperature change caused only by the energy deposition itself is denoted as ΔT_T^{dep}.

- The current change through the input coil of the SQUID ΔI_S is transformed (chapter 4.2.9) to the measured voltage output ΔU_{out}, where Φ_0 is the magnetic flux quantum:

$$\Delta U_{out} = \frac{\partial U}{\partial \Phi_0} \frac{\partial \Phi_0}{\partial I} \cdot \Delta I_S$$

- Finally, a calibration measurement defines the calibration factor c (chapter 4.2.10), which fixes the relation between the mean measured voltage output and the mean measured energy:[4]

$$E = c \cdot \Delta U_{out}$$

Summing all this up, the expected value of measured energy E after an energy deposition E_{dep} in the target is given by the following parameters:[5]

$$E = c \cdot \underbrace{\frac{\partial U}{\partial \Phi_0} \frac{\partial \Phi_0}{\partial I} \cdot \underbrace{\frac{I_T}{R_S + R_T} \cdot \underbrace{m \cdot \underbrace{\frac{r}{C_T} \cdot q\,p\,E_{dep}}_{E_{LD}}}_{\Delta T_T^{dep}}}_{\Delta R_T}}_{\Delta I_S}}_{\Delta U_{out}} \qquad (4.3)$$

4.2.1 Energy Transformation: p

The first step of a signal formation is the energy deposition in the target. Charged particles like alphas, betas, or muons distribute energy **directly** onto the components of the target (e$^-$, O, Ca, and W) via the electro-magnetic interaction. Uncharged particles, on the other hand, like gammas, neutrons, and WIMPs transfer energy **indirectly** onto the components of the CaWO$_4$ target. First, they transfer energy onto one single component of the target, which then itself distributes this energy onto the other target components via the electro-magnetic interaction. However, in both cases the energy is transferred onto the different components of the crystal at some point. If, for example, an energy of 10 keV is transferred onto an outer electron of an atom, this electron will lose its energy on its about 800 nm long path in the target mainly via collisions with other electrons [33]. The secondary electrons themselves lose their energies via further

---

[4] The larger the mean measured voltage output ΔU_{out} is, as smaller is the calibration factor c. For a constant measurement uncertainty (see section 4.3) a smaller calibration factor is equal to a better energy resolution.

[5] For a clear arrangement in this section expectation values are **not** marked by brackets, e.g. $E \equiv \langle E \rangle$.

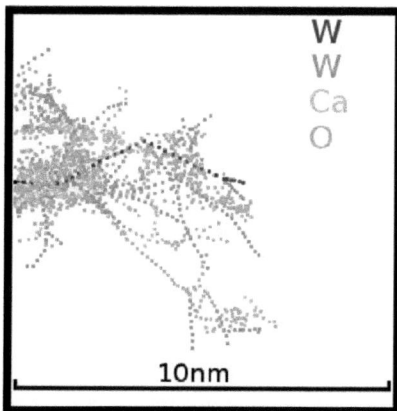

Figure 4.4: In this picture a typical simulated path of a tungsten nucleus with an initial energy of 10 keV in CaWO$_4$ is shown. On this path the nucleus loses energy via collisions with further components of the crystal. The paths of the secondary nuclear recoils on which they lose their energies again are also plotted.

collisions. If, on the other hand, the initial 10 keV are transferred onto a single tungsten nucleus, this nucleus will lose this energy on its only about 7 nm long path mainly via collisions with other nuclei [32]. A typical simulated energy loss of such a tungsten nucleus in a CaWO$_4$ target can be seen in figure 4.4. In this simulation a tungsten nucleus with an energy of 10 keV is entering a CaWO$_4$ crystal at the position of the black marker on the left hand side. The path of the initial tungsten nucleus in the target crystal is visualized in black. Secondary components, which took a part of the initial energy via a collision, can move in the target. Their paths are marked in their corresponding colors (W, Ca, and O). These components lose themselves their energies via further collisions on their marked paths in the crystal. In all these collisions phonons, vacancies, and **electronic excitations** can be created.

Generally, only energy transformed into electronic exictations can be converted into scintillation light, but it can also be converted into phonons or stay in trapped charged particles. Finally, most of the initially **deposited energy** is **converted** into **phonons**, while only a few percent are in form of **scintillating light**, cp. chapter 3.

4.2.2 Photon Transport: q

The second step towards a light channel signal is the photon transport. Created scintillation light can escape the crystal or be self-absorbed by the crystal as long as it is in its volume. The self-absorption depends on the self-absorption probability per path length and on the total path lengths of the photons in the crystal. While the self-absorption probability depends on the **quality of the crystal**, the path length depends on the **transmission probability**, **size**, and **shape** of the crystal. These four parameters will be commented next:

- The influence of the crystal's quality on the absolute light output can make a difference of more than a factor of two [36]. The **quality of a crystal** depends on the producer's skills and is given by the number of scintillation light producing centers and the crystal's self-absorption probability.

- To get an idea about the **transmission probability** from inside the crystal to the outside vacuum, the angle dependent probability can be derived with Snell's law and the Fresnel equations. The refraction index of calcium tungstate is $n_{CaWO_4} = 1.92$; for the outer vacuum it is $n_{vac} = 1$. The middle curve of figure 4.5 shows the derived transmission probability out of the crystal as a function of the angle of incidence. Total reflection is given for angles of incidence larger than 31°.
  Simplified, light with an angle of incidence smaller than 30° is transmitted, whereas larger angles are reflected.

- **Size** and **shape** of the CRESST-II crystals are given as cylindric with a diameter and height of 40 mm. Thus the crystal is a highly symmetric body, where recurrent incident angles in subsequent reflections appear. Hence light trapping in the crystal is reduced efficiently by roughening of one of the front sides, as mentioned above.

Above items are influenced by further parameters. Three of them should be mentioned here:

- The probability for **self-absorption by** the target crystal is increased due to the **thermometer film**, which measures the target temperature. It is evaporated on one of the front sides of the target with a size of 6 mm × 8 mm. The probability for light absorption due to the thermometer was determined to be ≈ 8 % [36].

- The path length in the crystal can be extended due to **re-entrance** of the light. Since crystal and light absorber are surrounded by a highly reflecting foil, escaped light can re-enter the crystal after reflection.

- In general, size, shape, and transmission probability of the crystal determine also the **area and direction of escape**.

Escaped scintillation photons can be reflected as long as they are not absorbed, cp. section 4.1.2. Scintillation light absorbed by the light absorber contributes to the detected light signal. The fraction of the energy deposited in the detector target which is absorbed by the light detector is typically in CRESST-II in the range of [35]:

$$1.0\,\% \lesssim pq \lesssim 2.5\,\%$$

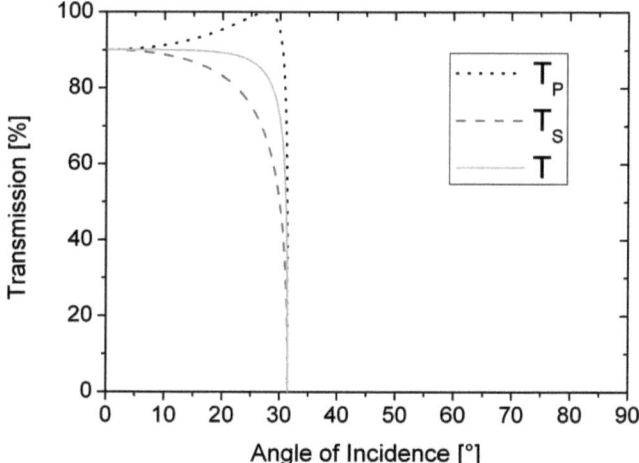

Figure 4.5: In this plot the transmission probability of light from CaWO$_4$ into vacuum is shown for all different angles of incidence. In light gray the value for non-polarized light is plotted. Black and gray show the probabilities for polarized light in (T_P) and perpendicular (T_S) to the plane which is spanned by the incident, reflected, and transmitted light.

4.2.3 Phonon Transport: r

The scintillation photons are absorbed via the photoelectric effect. In this process the photon energy of $E_{Photon} \approx 3$ eV is transferred on a single electron, which itself loses this energy in the light absorber by creating phonons of about half of the Debye energy. These phonons decay rapidly into still non-thermal phonons with an energy of $E_{Phonon} \approx 2.5$ meV $\hat{=}$ 25 K. At this energy the life time of the non-thermal phonons is long enough to distribute all over the absorber and thermometer [39]. Till the phonons decay to this energy of about 2.5 meV, the number of energy carriers has been increased by more than three orders of magnitude by this process. As a consequence of this, all following processes happen under high statistics. Therefore, the influence of energy quantization is negligible.

For the light signal it is of importance which amount of energy is **transported via phonons** into the thermometer's electrons.[6] For this energy transport three possibilities exist:

- First, the **direct absorption of non-thermal phonons** in the thermometer. The non-thermal phonons can transmit into the thermometer. Since

---

[6]It is assumed that the electron system dominates the influence on the measured electrical resistance of the thermometer. Hence the electron temperature of the thermometer is detected. This is confirmed by the measured pulse shape which fits to the results of the model that describes the thermometer's electron temperature, see chapter 5.

the mean free path of non-thermal phonons in the thermometer is expected to be smaller than the thermometer's thickness (200 nm) [40],[7] the absorption probability of transmitted phonons is high. Therefore, for this process the area of the thermometer is of importance. The larger the **thermometer area** is, the more non-thermal phonons will be absorbed.

- Second, non-thermal phonons are absorbed by **Cooper pairs** in the superconducting aluminum films, which are next to the thermometer, see figure 4.3(a). The mean free path of non-thermal phonons in these films is determined in [40] to be 1.8 kÅ, which is smaller than the bidirectional path through the films with a thickness of 1 kÅ. After absorption of a phonon the Cooper pair breaks up; broken Cooper pairs are called quasi particles. They can break up further Cooper pairs by emitting phonons and can **diffuse** into the thermometer and transfer energy onto the electrons of the thermometer.

- Third, non-thermal phonons can thermalize in the light absorber mainly due to surface contaminations. For this reason a third process is taking place. This process is the suppressed phonon-electron interaction of thermal phonons in the thermometer. This interaction is much weaker than the one of non-thermal phonons, since it is $\sim T_{\text{Phonon}}^5$ [41]. For this reason the mean free path length of thermal phonons in the thermometer is much larger than the thermometer thickness. Hence for the **absorption of thermal phonons** in the thermometer the **volume** is the relevant quantity. However, since thermal phonons are stable, also in this third way energy is transferred into the thermometer.

These three ways of energy transport into the electron system of the thermometer are possible and determine the fraction r of the absorbed energy which is transported into the thermometer.

Its value can be determined as follows: A typical measured temperature rise of the thermometer at the top of the transition, cp. next section, at a temperature of $T_T \approx 35\,\text{mK}$ is about $130\,\mu\text{K}$ for an energy deposition of 1 keV in the light absorber. With the help of the next section the fraction r can be derived:

$$r \approx 8\,\%$$

This value can be confirmed roughly by [28]. There, instead, the fraction ε^* of non-thermal phonons is estimated, which are absorbed directly by the thermometer. This fraction does not differ very much from r taking into account that the light channel signal is dominated by the fast component, cp. chapter 5. The determined value is in agreement to the value presented above:

$$r \approx \varepsilon^* \approx 5\,\%$$

---

[7]For a precise theoretical derivation of this process the behavior of longitudinal and transversal phonons has to be distinguished. For simplicity this will not be taken into account; in addition, theory and experiment can differ significantly from each other in this point.

4.2 Mean Measured Energy

A detailed discussion of the phonon transport in the light absorber will be presented with the help of a model in chapter 5.

4.2.4 Heat Capacity: C_T

The energy transferred into the electron system of the thermometer is distributed afterwards over the electrons due to the relatively strong electron-electron interaction, see section 5.3.2. The thermometer electron temperature increase is inversely proportional to its **heat capacity** C_T.

The heat capacity of the thermometer can be described in three different temperature ranges:[8]

- $T > T_c$: Above the transition temperature T_c the heat capacity of the light detector thermometer is given by, see appendix A:

$$C_T(T) = 17.86 \, \frac{\text{eV}}{\text{mK}} \cdot \frac{T}{\text{mK}} \qquad (4.4)$$

- $T = T_c$: At the transition temperature to the superconducting state there occurs a phase transition of second order. Therefore, the heat capacity has a discontinuity at T_c, see appendix A:

$$\Delta C_T(T_c) = 1.43 \cdot C_T(T_c)$$

- $T \ll T_c$: Significantly below the transition temperature the heat capacity can be described by, see appendix A:

$$C_T(T) \sim exp\left(-D \cdot \frac{T_c}{T}\right)$$

where D is a constant connected with the energy gap of the Cooper pairs.

The jump of the heat capacity at the transition temperature predicted by theory is confirmed experimentally, e.g. [42].[9] However, in practice the transition temperature is slightly dependent on the position, which causes a finite width of the transition, see figure 2.5.

Assuming that the geometric sizes of the TES film, the width b and the thickness d, are both not larger than the dimension on which the transition temperature changes and using a strictly linear change of the resistance depending on the transition temperature, then the temperature dependent heat capacity can be described as:

$$C_T(T_c^l \leq T \leq T_c^u) = \left(1.43 \, \frac{T_c^u - T}{T_c^u - T_c^l} + 1\right) \cdot C_T^u \qquad (4.5)$$

---

[8] A description of the temperature dependent heat capacities of non-metals, metals, and superconductors, and their respective CRESST-II specific values can be found in appendix A.
[9] The jump height of $\Delta C = 1.43 \cdot C(T_c)$ following from BCS theory is valid for superconductors with weak coupling between the electrons and the lattice as it is the case for tungsten.

Here is T_c^l the lower and T_c^u the upper end temperature of the transition. C_T^u is the thermometer heat capacity at the upper end temperature, which is equal to the heat capacity in the normal conducting state at T_c.

In practice the influence of the TES width b cannot be neglected and also a linear critical temperature change is not usual. For example, a significant part of the thermometer has to be normal-conducting to measure a finite thermometer resistance. For this reason heat capacity and resistance <u>cannot</u> have a linear relation. Therefore, the approximation shown above can only be a rough parameterization.

However, the heat capacity depends on the stabilized thermometer **temperature**. This point in the transition is called the **operating point**.

From equation (4.4) one can expect that the heat capacity is proportional to the absolute **transition temperature** T_c. This temperature can be influenced by e.g. stress or contaminations.[10]

Heat capacities are proportional to the **volume** V of the thermometer:

$$C_T = c_T^{specific} \cdot \varrho \cdot \underbrace{b \cdot d \cdot l}_{V}$$

Here is $c_T^{specific}$ the specific heat of the thermometer, ϱ is its density, and l is its length.

A typical value of the light detector thermometer heat capacity in CRESST-II is in the range of:

$$200 \, \frac{\text{eV}}{\text{mK}} \lesssim C_T \lesssim 1000 \, \frac{\text{eV}}{\text{mK}}$$

4.2.5 Transition Slope: m

The energy absorption in form of phonons in the thermometer and its heat capacity define the temperature rise ΔT_T^{dep} caused by the energy deposition in the target. The next step in the signal evolution is the resistance change. It is of interest how much the thermometer resistance changes (ΔR_T) for a temperature rise ΔT_T^{dep}. The connection between these two values is the **transition slope m**:

$$\Delta R_T = m \cdot \Delta T_T^{dep}$$

The transition slope depends on different parameters, which will be presented next.

In appendix C.3 is shown that the temperature rise of a thermometer ΔT_T^{dep} caused <u>only</u> by an energy deposition is similar to a temperature change of the bath temperature T_B:

$$\Delta T_T^{dep} \approx \Delta T_B$$

---

[10] A side effect of a lower transition temperature is that the coupling of the phonons thermalized in the absorber with the electrons of the thermometer is weakened.

4.2 Mean Measured Energy

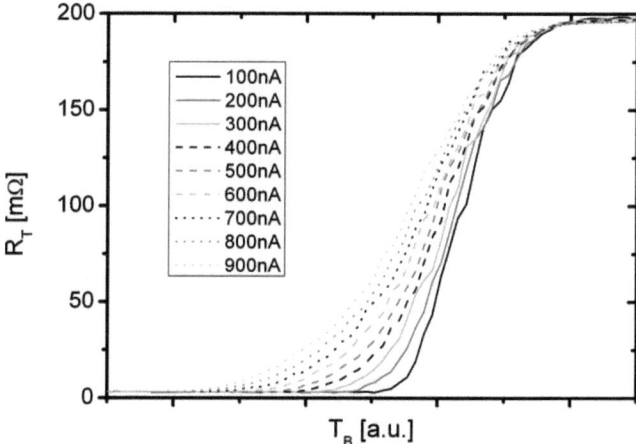

Figure 4.6: In the shown measurement a light detector thermometer transition is measured as a function of the bath temperature for nine different total bias currents. A difference in the bath temperature ΔT_B is similar to a temperature difference ΔT_T^{dep} caused by an energy deposition in the target. Both are not influenced by current effects, whereas the measured resistance depends additionally on the heating due to the bias current and critical bias effects. For this reason the observed transition slopes depend on the bias current.

For this reason the transition slope m describes the relation between a resistance change and a bath temperature change, too:

$$\Delta R_T \approx m \cdot \Delta T_B$$

In figure 4.6 this relation for the transition slope is shown for different **bias currents** I_{Tot}. In this plot the thermometer resistance is measured for different bath temperatures and for the bias currents $I_{Tot} = 100, 200, ..., 900$ nA. It can be seen that a small residual resistance remains below the transition. Such an effect can appear due to contact resistances in the circuit branch of the thermometer, see figure 4.3(b). The normal conducting thermometer resistance is about 200 mΩ, which is in the usual range of 100 - 300 mΩ.

One can see that the transition slope depends on the bias current. Reasons for this are **critical current effects** and **self-heating effects** due to the bias current read out of the thermometer resistance.

Also with negligible current effects a transition has a finite width, and therefore a finite slope. The reason for this effect is the **position dependent transition temperature** T_c, which is caused for example by stress, defects, or contaminations in the lattice of the thermometer crystal.

Typical values for the transition slope are in the range:[11]

$$100\,\frac{\mathrm{m}\Omega}{\mathrm{mK}} \lesssim m \lesssim 3\,000\,\frac{\mathrm{m}\Omega}{\mathrm{mK}}$$

4.2.6 Thermometer Bias Current: I_T

The resistance change of the thermometer is read out via the **current through the thermometer** I_T. The signal is proportional to the bias current change, cp. equation (4.3). For this reason, high currents are preferred since the measurement uncertainty shows only a weak dependence on the bias current, cp. section 4.3. For an upper limit of this thermometer bias current two effects have to be taken into account:

- **Self-heating effects**: The thermometer current dissipates a power in the thermometer system: $P_{Bias} = R_T \cdot I_T^2$. This energy has to be dissipated via the weak thermal coupling to the bath. This energy flow limits the usable thermometer current.

- **Critical current effects**: Superconductivity breaks down when the current is larger than the critical current. In general, critical currents are always large compared to the thermometer currents used here. However, these thermometers are thermally stabilized in the transition where the superconductor gets normal conductive. In this range the critical current comes down to zero. Therefore, this effect has to be taken into account. A limitation of the thermometer current due to critical current effects shows up in the transition slope discussed above.

For reading out a light channel thermometer, typical currents in the range

$$0.1\,\mathrm{\mu A} \lesssim I_T \lesssim 1\,\mathrm{\mu A}$$

are used.

Practical Hint: The total bias current I_{Tot} is set up instead of the thermometer bias current I_T. Therefore, it is easier to determine the dependency of the mean measured energy on this value, cp. equation (4.3):

$$\frac{I_T}{R_S + R_T} = \frac{I_{Tot}\,R_S}{(R_S + R_T)^2} \qquad (4.6)$$

This formulation points the strong dependency of the mean measured value on R_T (for $R_T \gtrsim R_S$) out. However, the thermometer resistance increases also due to self-heating and critical current effects for increasing bias currents: $R_T(I_T; I_{Tot})$

---
[11]It should be noted that, for Dark Matter detection, due to the small amount of energy transfer only the **local transition slope** is of importance. The steepest part of a transition can thus be chosen.

4.2.7 Shunt Resistance: R_S

The shunt resistance is placed in parallel to the thermometer resistance, see figure 4.3(b). It determines the splitting ratio of the total bias current into the two parallel branches.

From equation (4.3) one can see that its influence on the measured energy is **little** as long as it is smaller than the resistance of the thermometer as it is usually the case in CRESST-II: $R_S < R_T$.

In CRESST-II the shunt resistance is:

$$R_S = 40\,\text{m}\Omega$$

4.2.8 Thermometer Resistance: R_T

R_T is the resistance of the thermometer. It depends on the operating point, i.e. the thermally stabilized point in the transition; therefore it can be between zero and the normal conducting value of the thermometer. It can be seen from the equations (4.3) and (4.6) that the **measured signal** increases for smaller thermometer resistances.

In figure 4.7 the dependence of the **self-heating** ($P_{Bias} = R_T \cdot I_T^2$) of the bias current of the thermometer resistance can be seen. The self-heating is maximal for $R_T = R_S$. For typical values ($I_{Tot} = 1\,\mu\text{A}$; $R_S = 40\,\text{m}\Omega$) the maximum heating is $P_{Bias} = 10\,\text{fW}$.

In CRESST-II the thermometer resistance it usually in the range:

$$50\,\text{m}\Omega \lesssim R_T \lesssim 300\,\text{m}\Omega$$

4.2.9 Current to Voltage Transformation: $\frac{\partial U}{\partial \Phi_0}\frac{\partial \Phi_0}{\partial I}$

The change of the thermometer resistance causes a change in the bias current splitting, since the total current I_{Tot} is constant, see figure 4.3(b). The corresponding bias current change is detected via a **SQUID**.[12] In the schematic drawing 4.3(b) only the input coil, which induces a magnetic flux in the SQUID, is shown. When the current changes, also the induced magnetic field in the SQUID changes. Via an additional feedback coil and a feedback control the magnetic flux is held constant in the SQUID, i.e. for a change in the current splitting the current through the feedback coil changes to compensate the change of the magnetic flux through the SQUID. The voltage controlling this current is the **signal output**.

The SQUID converts the current change into a voltage output. The feedback control is adjusted such that the current-to-voltage relation is linear. The proportionality constant is affected by the geometry of the SQUID coil. However, the

---
[12] The SQUID is used as pre-amplifier.

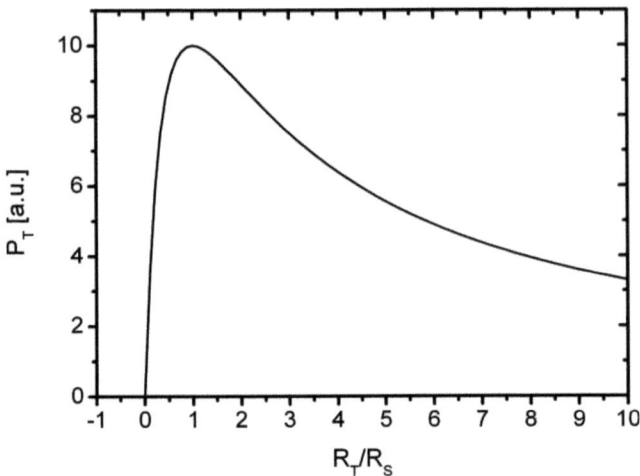

Figure 4.7: In this figure the heating power of the thermometer bias current $P_{Bias} = R_T I_T^2$ is plotted for different thermometer resistances R_T. The total bias current I_{Tot} is constant. It can be seen that the heating effect is maximal for $R_T = R_S$. In CRESST-II $R_S = 40\,\mathrm{m}\Omega$. For a typical total bias current $I_{Tot} = 1\,\mathrm{\mu A}$ the maximal bias heating in the thermometer is $P_{Bias} = 10\,\mathrm{fW}$.

signal gain due to this conversion counts in the same way for the measurement fluctuations, arisen up to this step of signal evolution.

Conversion factors in CRESST-II are typically:

$$\frac{\partial I}{\partial \Phi_0} = 200\,\frac{\mathrm{nA}}{\Phi_0} \quad \text{and} \quad \frac{\partial U}{\partial \Phi_0} = 10\,\frac{\mathrm{V}}{\Phi_0}$$

$$\Rightarrow \quad \frac{\partial U}{\partial \Phi_0}\frac{\partial \Phi_0}{\partial I} = 50\,\frac{\mathrm{mV}}{\mathrm{nA}}$$

4.2.10 Energy Calibration: c

As last step of the light signal evolution an energy calibration determines the relation between measured voltage output ΔU_{out} and mean measured energy E. For calibration this measured energy is set equal to the known deposited energy E_{dep}, which defines the energy calibration factor c:

$$E \equiv E_{dep}$$

$$\Rightarrow \quad c := \left(\frac{\partial U}{\partial \Phi_0}\frac{\partial \Phi_0}{\partial I} \cdot \frac{I_T}{R_S + R_T} \cdot m \cdot \frac{r}{C_T} \cdot pq\right)^{-1}$$

4.3 Measurement Uncertainty

Besides the mean measured energy (section 4.2), the measurement uncertainty in the light channel influences mainly the widths of the bands in the light-phonon-plane, see beginning of this chapter. The **uncertainty of the light channel signal** will be discussed in this section.

Since systematic uncertainties are mainly ruled out by the energy calibration factor c, uncertainties of a light signal measurement in CRESST-II are mostly of statistical nature. Reasons for these uncertainties are differences in the signal evolution caused by the energy quantization, noise, as well as picked up interferences adding to the signal.
In general, in each step of signal evolution information can only be lost by these influences.

∥ **During signal evolution information can only be lost.** ∥

This means that any information loss at the beginning of the signal evolution cannot be compensated by a subsequent step. For example, a fluctuation in the transformation of the deposited energy into scintillation light cannot be compensated by a steeper transition slope afterwards. For this reason it has to be taken into account **at which step which measurement uncertainty takes place**. In the following the most important uncertainties will be discussed in the order of the signal evolution in the light channel after an energy deposition E_{dep} in the detector target.

4.3.1 Energy Transformation: δp

After a fixed energy deposition in the detector target a fraction of this energy is converted into scintillation light. This fraction is distributed statistically around its mean value: $p = \langle p \rangle \pm \delta p$. The distribution depends on two parameters. First, on the **number of created photons**. This number is strongly correlated to the path length of the primary and secondary recoils in the target, cp. chapter 3. Second, on the **energies of the individual photons**. The energy distribution of the photons can be seen from the emission spectrum of figure 4.2. The effect of the emission spectrum is much smaller than the fluctuation of the photon number due to the narrow energy distribution.

4.3.2 Photon Transport: δq

The fluctuation of the photon fraction absorbed by the light absorber is determined by two effects: First, the statistical fluctuation due to the **probabilities of transmission and absorption**. Second, the **position dependence** in the crystal: Depending on the position in the crystal where light is produced the probability of absorption by the light absorber changes. For example, the probability for light produced is minimal very close to the crystal thermometer, due to the possible absorption by the thermometer.[13] However, the influence of the local dependency is small in the energy range of interest [28].

4.3.3 Absorbed Energy: δE_{LD}

Up to now it is not known whether the **number of created photons p** or the **fraction of absorbed photons q** or both effects dominate the uncertainty of absorbed light δE_{LD}. However, it is known that the total energy resolution of the light channel is **dominated** by the uncertainty of absorbed energy δE_{LD} for energies $E \gtrsim 5\,\text{keV}_{ee}$ [28].

4.3.4 Phonon Transport: δr

The fluctuation of the number of phonons, which are transported to the thermometer and heat it up, and their corresponding energy distribution is determined by the **phonon transmission and absorption probability**, the **phonon decay rate**, and the **position dependence in the light absorber**.

The influence of the **position dependence** on the measurement uncertainty was measured in [38]. Only for events directly behind the thermometer structure an signal increase of about 7 % was observed. Elsewhere no position dependence was observed. Due to this small effect and the small area of the thermometer structure the position dependence of the light absorber can be **neglected**.
As seen in section 4.2.3, the energy of the phonons is orders of magnitudes smaller than the energy of a scintillation photon:

$$\langle E_{Phonon} \rangle \approx 2.5\,\text{meV} \quad \ll \quad \langle E_{Photon} \rangle \approx 3\,\text{eV}$$

For this reason the large number of phonons causes a statistically stable energy flow into the thermometer. The **energy quantization** and all related effects influencing the measurement uncertainty, such as the phonon transmission and absorption probability, and the phonon decay rate, can be **neglected** and do not contribute significantly to the measurement uncertainty.

4.3.5 Heat Capacity: δC_T

The uncertainty of the heat capacity is given by the **stability of the operating point** in the transition. A proportional–integral–derivative controller (PID controller) stabilizes the thermometer thermally at the operating point in the transition. The thereby given temperature uncertainty is reflected in the heat capacity uncertainty δC_T. Due to the precise, long time stability of the operating point [43], it is expected that this effect is **negligible**.

---

[13] A new composite technique [35], where two crystals are glued together, is expected to reduce this effect of light absorption by the target thermometer.

4.3 Measurement Uncertainty

4.3.6 Thermometer Temperature Rise: $\delta(\Delta T_T^{dep})$

Beside all previous effects the variation of the thermometer temperature rise is additionally determined by the **phonon noise** and the pickup of **electrical interferences**.

Phonon noise, also called thermal noise, are phonon fluctuations between the thermometer and the connected systems. Due to the random movement of phonons, the thermometer temperature cannot be absolutely constant.

In section 5.3.2 the weakness of the electron-phonon coupling at ultra-low temperatures is described. Due to this weak coupling of the phonons to the measured electron temperature in the thermometer the influence of this effect onto the measurement uncertainty is expected to be small. In [28] the influence of the phonon noise is determined to be **negligible**.

Electrical interferences can be picked up by the read out and control feed lines of the thermometer structure. These are four wires where two wires are connected to the thermometer heater to stabilize it thermally, and two wires are connected to read out the thermometer's resistance. In figure 4.3(a), beside these four wires, an additional cooling wire can be seen, too. In figure 4.8 a schematic drawing of these four wires and the main resistances of the signal measurement are shown. Each of the two wire pairs is twisted to reduce the **differential noise**. Between the two pairs a **common mode noise** can be picked up, because there is an electrical connection in between. This kind of interference is not reduced by twisted pair wires. The differential and the common mode noise can dissipate heat in the different resistances of the thermometer structure.[14] This way the temperature of the thermometer can be changed. On the left hand side of figure 4.9 the influence of picked up 100 Hz interferences on a light detector signal can be seen. The long decay time, which is as long as the decay time of the detected pulse, indicates a real heat dissipation due to the interferences. By careful design and additional shielding the influence of the electrical interferences can be reduced to a **negligible** value.

4.3.7 Transition Slope: δm

The uncertainty of the transition slope is given by the **stability of the operating point** in the transition, too. In addition to the thermally stable thermometer temperature, cp. section 4.3.5, the transition slope is usually very insensitive on the operating point due to the linearity of the transition. Due to these two arguments this uncertainty can be **neglected**.

4.3.8 Thermometer Resistance Change: $\delta(\Delta R_T)$

Resistances are <u>not homogeneous</u> on microscopic scale; they are dependent on the position on the film. Local resistance fluctuations seem to be connected to the presence of **1/f-noise** [44].

---

[14]The effect of interferences directly coupled into the SQUID input coil are discussed in section 4.3.9.

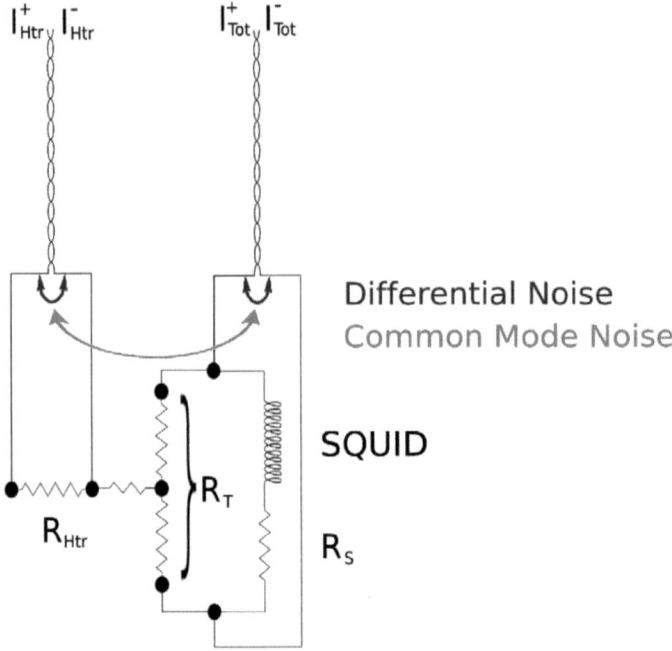

Figure 4.8: In this figure a schematic drawing of the thermometer structure and its feed lines are shown. The bias current to control the thermometer temperature, I_{Htr}^+ and I_{Htr}^-, is introduced via twisted pair wires which are connected to the heater resistance R_T to reduce the **Differential noise**. In the same way the thermometer read out bias current, I_{Tot}^+ and I_{Tot}^-, is introduced via twisted pair wires. These wires are connected to the two parallel branches, the SQUID and shunt resistor and the thermometer branch. Due to the used strong connection of heater and thermometer, there exists also an electrical connection between these two parts. Via this connection **Common Mode noise** can be introduced between one of the thermometer and one of the heater wires. All kinds of electrical interferences can dissipate heat into the thermometer system. On the left hand side of figure 4.9 the effect of 100 Hz introduced noise on a measured pulse can be seen. Each increase of the measured thermometer temperature cools down with a finite time constant. The time constant is equal to the decay time of the measured pulse, i.e. the thermometer is physically heated up. On the right hand side, on the contrary, the influence of interferences are shown which are introduced directly into the SQUID of a target thermometer. Here the decay time is much smaller. The thermometer is not heated up by these interferences. This effect is discussed in section 4.3.9.

4.3 Measurement Uncertainty

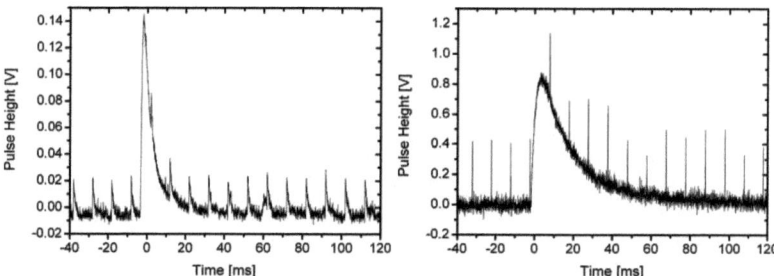

Figure 4.9: Both pictures show measured thermometer pulses overlapped by 100 Hz interferences. On the left hand side, a light detector pulse is shown with interferences which dissipate heat in the thermometer system. This can be seen from the heat-like decay time of the signal after each 10 ms. On the right hand side, a phonon detector pulse is shown. In this case the interferences decay immediately. These are interferences coupled directly into the SQUID input coil.

Locally varying resistances can have their origin, for example, in the motion of defects. In [45] it could be shown that induced defects in thin copper films increase the $1/f$-noise. On the other hand, it was successfully shown in [46] that annealing, and therefore the reduction of the defects, reduces the $1/f$-noise.
Another reason for local differences of a resistance can be stress in the lattice. [47] describes that stress has a significant effect on the low-frequency excess noise of metal films, which is similar to the $1/f$-noise. Strain relaxations are assumed to introduce much of the their observed noise.
In CRESST-II transition edge sensors (TES) are used. Given the finite transition width, the thermometer resistance must be dependent on position. It is expected that this effect introduces $1/f$-noise into the system. In [28] a **significant** amount of $1/f$-noise has been determined based on the noise increase at low frequencies.

4.3.9 SQUID Bias Current Change: $\delta(\Delta I_S)$

The change of the SQUID bias current during a pulse is overlapped, beside all previous effects, directly by two main effects: **External interferences** and **Johnson-Nyquist-noise**.
External interferences can be picked up by the bias current lines, which are connected to the thermometer structure, see figure 4.8. When these interferences are coupled directly into the SQUID input coil this effect shows up as on the right hand side of figure 4.9. In this picture a target thermometer pulse is shown overlapped by 100 Hz interferences. The very short decay time of the interferences shows that no temperature rise of the thermometer structure appears. These interferences are only seen in the SQUID. By careful design, such interferences can be reduced down to a **negligible** level.
Johnson-Nyquist-noise is white noise created by electron fluctuations due to their thermal agitation, independent of an applied voltage. In the CRESST-II

readout circuit three different electronic resistances contribute mainly to this noise, which is coupled directly into the SQUID input coil. These are thermometer and shunt resistance, as well as the resistive shunts in parallel to the Josephson junctions in the SQUID. These two shunt resistances are **not** shown in figure 4.3(b). However, they are introduced together with also not shown parallel capacities to eliminate hysteresis effects of the SQUIDs [48][49]. These shunts contribute the dominating noise of the SQUIDs [50]. In [28] the contribution of the Johnson-Nyquist-noise caused by **thermometer** and the **SQUID shunts** onto the total noise amount is determined to be **significant**.

4.3.10 Summing Up and Conclusions

In the above sections, the main reasons for the measurement uncertainty of the energy signal have been itemized. In the following, equation (4.3) is shown again; the above uncertainties are added to this equation by arrows. **They are placed at the point of signal evolution when they show up**, although they have influence on all following steps of the signal development. For this reason the direction of the signal development is marked:[15]

$$E = c \cdot \frac{\partial U}{\partial \Phi_0} \frac{\partial \Phi_0}{\partial I} \cdot \frac{I_T}{R_S + R_T} \cdot m \cdot \frac{r}{C_T} \cdot q\, p\, E_{dep}$$

with arrows indicating δm, δr, δq, δp, δC_T, and underbraces:
E_{LD}
$\Delta T_T^{dep} \Leftarrow \delta(\Delta T_T^{dep})$
$\Delta R_T \Leftarrow \delta(\Delta R_T)$
$\Delta I_S \Leftarrow \delta(\Delta I_S)$
ΔU_{out}

The different effects (uncertainties) contributing to the measurement uncertainty are summarized in the following by **highlighting** those with the **strongest influence**:

δp	**Number of created photons** Energy distribution of the individual photons
δq	**Probabilities of transmission and absorption** Position dependence of the target
δr	Phonon transmission and absorption probability Phonon decay rate

---

[15] For a clear arrangement in this equation expectation values are **not** marked by brackets, e.g. $E \equiv \langle E \rangle$.

4.3 Measurement Uncertainty

	Position dependence of the light absorber
δC_T	Stability of the operating point
$\delta(\Delta T_T^{dep})$	Phonon noise
	Electrical interferences
δm	Stability of the operating point
$\delta(\Delta R_T)$	**1/f-noise**
$\delta(\Delta I_S)$	**Johnson-Nyquist-noise**
	External interferences

The four **marked** sources mainly induce the measurement uncertainty of the CRESST-II light channel. To determine their relevance for the WIMP identification threshold, these causes can be divided into **energy-dependent** and **energy-independent**.

In [28] the resolution of a typical light channel has been fitted by an energy-dependent and an energy-independent part. With the help of this analysis one can see that the measurement uncertainty of the light channel is dominated by the energy-independent measurement uncertainties below $\approx 5\,\text{keV}_{ee}$. The width of the electron band is thus dominated by the energy-independent part below $\approx 5\,\text{keV}$ of energy deposition in the detector target and the width of the tungsten band below $\approx 200\,\text{keV}$, due to the quenching factor.

The same amount of energy-independent uncertainty is confirmed by the analysis of the heater pulse height [28]:[16] The distribution of the thermometer temperature change caused by heater pulses represents the energy resolution for zero energy, which is also called **baseline noise**. The baseline noise is dominated by the **Johnson-Nyquist-noise caused by thermometer and SQUID**, and the **1/f-noise introduced by the thermometer** [28].

For the WIMP sensitivity of the CRESST-II experiment the overlap of the background bands with the tungsten-band is of interest. The overlap is determined by the energy resolution of the light channel. In addition to the baseline noise the resolution in the relevant energy range is given by the parameters of signal evolution which create the mean measured energy before the baseline noise sources appear in the signal development. As seen, these are: $\boldsymbol{p, q, r, C_T, m, I_T, R_S}$, and $\boldsymbol{R_T}$.

> **The WIMP identification threshold can be lowered significantly by a reduction of the baseline noise or an increase of values which contribute to the signal before the baseline noise sources appear.**

---

[16]Heater pulses are injected energy pulses into the thermometer for testing the stability of the thermometer by reading out its temperature change due to these injected pulses.

Chapter 5

Model of the Energy Transport in the Light Detector

The most complex part of the overall light channel is the behavior in the light detector itself. For research and development of the light channel it is absolutely necessary to understand this part, see previous chapter. Therefore, in this chapter, a model describing the physics in the light detector will be presented.

5.1 Introduction

After an energy deposition in the target in average the fraction $p \cdot q$ of this energy is transported to the light absorber in the form of scintillation light. This energy is named E_{LD}. The energy fraction of this energy transported into the thermometer is described in chapter 4 by r. This energy fraction increases the thermometer temperature as described in equation (4.1):

$$\Delta T_T^{dep} = \frac{r\, E_{LD}}{C_T}$$

The model of this chapter **connects** the energy fraction r, which is transported into the thermometer and warms it up, to the **physical properties** of the light detector, like geometry, temperature, or the different materials used. Additionally, the model describes the time dependent change of the thermometer temperature:[1]

$$\Delta T_T = \Delta T_T(t) = \Delta T_T(r(t))$$

The model starts with an energy deposition in the light absorber, as it is the case for scintillation light absorption. It describes the energy transport via phonons

---

[1] The thermometer temperature change affected only by the energy deposition is denoted as ΔT_T^{dep}. Additionally, the current bias has influence on the thermometer temperature change. The total thermometer temperature change is therefore denoted as ΔT_T.

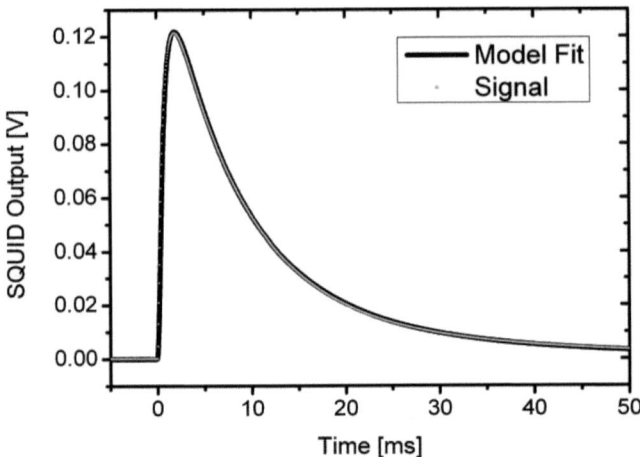

Figure 5.1: In this figure an averaged measured light channel pulse is shown in gray dots. The SQUID output voltage is proportional to the temperature change of the thermometer. The model introduced in this chapter predicts a time dependent temperature change of the thermometer after an energy deposition in the target. This time dependent temperature function depends on parameters which are fixed by fitting this function to the measured pulse. The fit is shown in black.

in the light detector itself, and the following time dependent temperature rise and cooling down phases of the thermometer. Although there are simplifications, it prognoses a time dependent thermometer temperature $\Delta T_T(t)$ which fits very well to the measured ones [39], see figure 5.1.[2]

5.2 Model Assumptions

▶ The light detector consists mainly of a non-metal light absorber and a metal thermometer. Both are thermally coupled to the heat bath. These constituent parts are shown schematically on the left hand side of figure 5.2. The first assumption of the model is, that those shown in the picture, the heat capacities, temperatures, and couplings, are the ones which dominate the behavior of $\Delta T_T(t)$.

---

[2]It should be noted that at the used ultra-low temperatures of about 15 mK the electron and the phonon system of the thermometer are only weakly coupled to each other [39]. For this reason these two systems can have different temperatures. It is assumed that out of these two the electron system dominates the measured electrical resistance of the thermometer. Hence the electron temperature of the thermometer is detected. This is confirmed by the measured pulse shape which fits to the results of the model that describes the thermometer electron temperature.

5.2 Model Assumptions

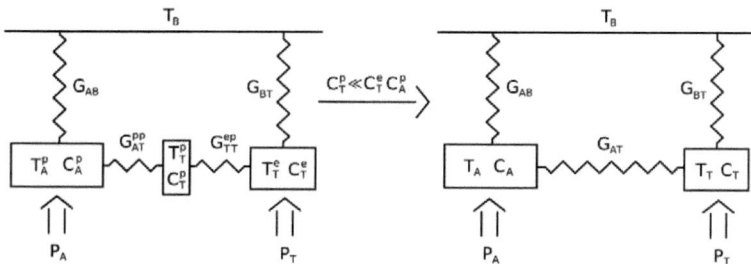

Figure 5.2: On the left hand side, main components of a light detector can be seen. These are the absorber and the thermometer which are characterized by their temperatures (T_A^x, T_T^x) and heat capacities (C_A^x, C_T^x). To take the thermal decoupling of the thermometer's electron and phonon system caused by the strong temperature dependence ($\sim T^5$, see section 5.3.2) into account, both of them are shown explicitly. Power inputs into absorber and thermometer are displayed by arrows (P_A, P_T). On the right hand side the influence of the thermometer phonon system C_T^p compared to the two other heat capacities is neglected. This assumption simplifies the system which is used for the model. The bath temperature T_B is constant due to its assumed infinitely large heat capacity C_B. The couplings between the different systems are denoted by G.

▶ A second assumption is that the phonon heat capacity of the thermometer C_T^p is negligibly small compared to the phonon heat capacity of the absorber C_A^p and to the electron heat capacity of the thermometer C_T^e:

$$C_T^p \ll C_A^p, C_T^e$$

This assumption is fulfilled for the light detector since at the operating temperature of about 15 mK C_T^p is at least five orders of magnitude smaller than C_A^p and C_T^e (cp. appendix A).

In this case the thermal coupling G_{AT} between the phonon temperature of the absorber T_A^p and the electron temperature of the thermometer T_T^e can be described as:

$$\frac{1}{G_{AT}} = \frac{1}{G_{AT}^{pp}} + \frac{1}{G_{TT}^{ep}}$$

G_{AT}^{pp} is the phonon-phonon (Kapitza) coupling between absorber and thermometer and G_{TT}^{ep} is the electron-phonon coupling in the thermometer itself, see figure 5.2. Under this assumption the description can be simplified:

$$\begin{aligned} C_A &:= C_A^p \\ C_T &:= C_T^e \\ T_A &:= T_A^p \\ T_T &:= T_T^e \end{aligned}$$

▶ However the main assumption of the model is:

‖ **There are only two energy states for phonons.** ‖

These are the non-thermal and the thermal state instead of a continuous energy state distribution. These two states are approximately the initial and the final states of phonons which are created after an energy deposition.[3] This assumption is well motivated since mainly phonons in these two states transfer energy into the thermometer electron system: As seen in section 4.2.3, due to the strong electron-phonon interaction of non-thermal phonons in the thermometer [39] **non-thermal phonons** transport energy efficiently to the thermometer electrons. Second, due to the stable thermal phonon state after decaying **thermal phonons** transfer, despite their long interaction time, energy into the thermometer, too. A two phonon state model is thus motivated.

In reality, the phonons are created in the silicon layer where the photons are absorbed. The frequencies of these phonons can be approximated by half of the Debye frequency of silicon: $\nu_D^{Si}/2 \approx 6.8\,\text{THz}$ [39]. Compared to the response time of the thermometer (180 μs - 410 μs, cp. section 5.5.5) their decay induced by lattice anharmonicities down to frequencies of about 300 GHz is fast ($\approx 100\,\mu$s), see figure 5.3. Their decay rate is strongly frequency dependent:

$$\tau_{decay} \sim \nu_{Phonon}^5$$

Compared to the response time of the thermometer these phonons are relatively stable and after about 20 μs they are distributed homogeneously all over the absorber.[4] In the model these are the initial **non-thermal** phonons. They are treated as an ideal gas. At this time, their anharmonic decay rate is negligible. Instead the rate for absorption by thermometer electrons and thermalization via inelastic scattering at the surfaces of the absorber dominates. Phonons thermalized in the absorber are described in the model as phonons in the **thermal** state.

---

[3] This is valid in the same way for phonons created after an energy deposition in the target crystal.
[4] The propagation direction after creation is connected with the density of states of the phonons. A phonon needs about 10 μs to cross the absorber once and for each of the three phonon modes exists a different angle of reflection.

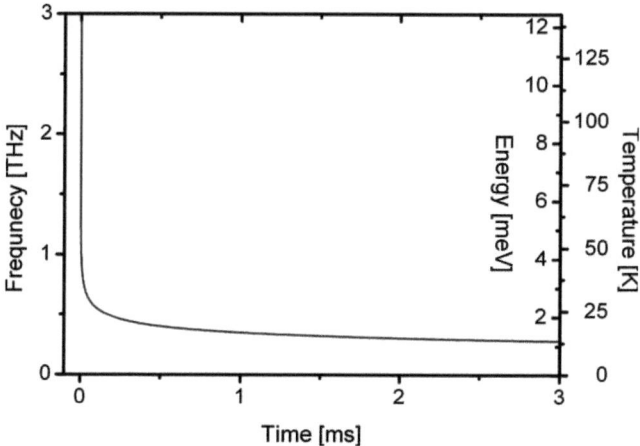

Figure 5.3: The derived average phonon frequency of non-thermal phonons in silicon depending on time. After an energy deposition non-thermal phonons with about half of the Debye frequency are created: $\nu_D^{Si} \approx 6.8\,\text{THz}$. The decay rate is caused by lattice anharmonicities and is proportional to ν_{Phonon}^5. Beside the phonon frequency also their proportional energy and temperature are indicated. The simulation is taken from [39].

5.3 Equations for the Temperatures

5.3.1 Basic Equations for the Temperatures

In this section the equations describing the thermometer temperature as a function of time will be presented. They describe the energy input via non-thermal phonons into the thermometer and the absorber, the following thermal adjustments of the subsystems via the thermalized phonons, and the relaxation of the light detector system back to the initial conditions.

The non-thermal phonons which are created in the absorber during an energy deposition can decay in the absorber, escape through the holding clamps, or be absorbed by the thermometer electrons. ε is the fraction of all non-thermal phonons which are absorbed in the thermometer. E_{LD} is the energy in form of scintillation light which is absorbed by the light detector. Then the thermometer is warmed up first by the energy $\varepsilon \cdot E_{LD}$ due to the non-thermal phonons.

The non-thermal phonons which decayed via inelastic scatterings in the absorber into thermal ones, warm up the absorber with the according energy $(1-\varepsilon) \cdot E_{LD}$.[5] Hence the time dependent power flows transferred by non-thermal phonons into

---

[5]In this model is assumed that all non-thermal phonons heat up absorber <u>or</u> thermometer. In reality a small fraction of non-thermal phonons warms up the light detector holding clamps and the cooling structure connecting the heat bath, too. In this way, a part of the non-thermal phonons escapes the system.

the thermometer and the absorber are given as:

$$P_T(t) = \varepsilon \cdot E_{LD} \cdot \frac{1}{\tau_N} e^{\left(-\frac{t}{\tau_N}\right)} \cdot \Theta(t) \tag{5.1}$$

$$P_A(t) = (1-\varepsilon) \cdot E_{LD} \cdot \frac{1}{\tau_N} e^{\left(-\frac{t}{\tau_N}\right)} \cdot \Theta(t) \tag{5.2}$$

Here, the $\Theta(t)$ function defines the point in time of the energy deposition in the absorber, i.e. the energy deposition is assumed to be infinitely fast.
τ_N is the life time of the non-thermal phonons. It depends on the degradation time τ_T caused by the electron absorptions in the thermometer and the degradation time τ_A caused by the decays in the absorber:

$$\frac{1}{\tau_N} = \frac{1}{\tau_T} + \frac{1}{\tau_A}$$

These two degradation times fix also the fraction ε of non-thermal phonons which are absorbed in the thermometer:

$$\varepsilon = \frac{\tau_A}{\tau_A + \tau_T}$$

After an energy deposition the thermal steady state of the system is broken. The time dependent temperatures $T_T(t)$ and $T_A(t)$ are changed first due to the power flows $P_T(t)$ and $P_A(t)$, respectively. The different thermal couplings afterwards adjust the temperatures.[6]

The two time dependent temperatures $T_T(t)$ and $T_A(t)$ can be described by the following coupled differential equations:

$$C_T \cdot \dot{T}_T + G_{AT} \cdot (T_T - T_A) + G_{BT} \cdot (T_T - T_B) = P_T$$
$$C_A \cdot \dot{T}_A + G_{AT} \cdot (T_A - T_T) + G_{AB} \cdot (T_A - T_B) = P_A$$

T_B is the constant bath temperature, G_{BT} and G_{AB} are the couplings between the bath and the thermometer and the absorber, respectively.
The power flows through the thermal couplings G_{AT}, G_{BT} and G_{AB} are assumed to be proportional to the respective temperature differences: $P = G \cdot \Delta T$. This is only valid for small temperature differences. For large temperature differences as it is usually the case, the coupling changes with the temperature. The thermal coupling finally depends on the bath temperature and the relative temperature difference, see next section.

---
[6]This can imply an even further temperature increase of the thermometer temperature before it relaxes back again to the steady state temperature, cp. section 5.5.2 and 5.5.3.

5.3.2 Thermometer Heater

During operation of the experiment many detectors are used at the same time. Typically their thermometers have slightly different operation temperatures T_T due to their different transition temperatures. The bath temperature T_B, on the other hand, is the same for all detectors, i.e. the bath temperature has to be at least as low as the lowest transition temperature so that all thermometers can be operated. Since for this reason the bath temperature is generally lower than the operation temperature of the thermometer, each thermometer is equipped with a **heater** which warms it up into transition again, see figure 4.3(a). The heater mainly warms up the individual thermometer with its needed $\boldsymbol{P_{Htr}}$, while the power input into the absorber can be neglected:

$$C_T \cdot \dot{T}_T + G_{AT} \cdot (T_T - T_A) + G_{BT} \cdot (T_T - T_B) = P_T + P_{Htr}$$
$$C_A \cdot \dot{T}_A + G_{AT} \cdot (T_A - T_T) + G_{AB} \cdot (T_A - T_B) = P_A$$

This constant additional power input P_{Htr} introduced due to the lowering of the bath temperature T_B has two side effects: A change of the absorber's heat capacity and a change of the thermal couplings.
The reason for these two effects is, that in the steady state ($P_T \equiv P_A \equiv \dot{T}_T \equiv \dot{T}_A \equiv 0$) the following **three temperatures are not equal anymore** due to the introduction of P_{Htr}:[7]

$$T_T^0 = T_A^0 = T_B \quad \Longrightarrow \quad T_T^0 > T_A^0 > T_B$$

If the time independent bath temperature T_B is lowered, the thermometer has therefore to be heated up again so that it has the same temperature as before, cp. first differential equation. The absorber temperature T_A^0 adjusts depending on the couplings in between, cp. second differential equation. The consequences for the physical system are:

- First, a lower absorber temperature induces a lower **absorber heat capacity**, cp. appendix A:

$$C_A \sim T_A^3$$

- Second, the **thermal couplings change**. For small temperature differences the power flow can be described as $P = G \cdot \Delta T$. In this used description the thermal coupling depends directly or indirectly on the bath temperature T_B. This will be shown in the following.

  In general the power flow through a thermal coupling between two systems at low temperatures T_1 and T_2 can be written as:

  $$P = \Pi \cdot (T_1^n - T_2^n) \tag{5.3}$$

---
[7] $T^0 \equiv T(t \leq 0)$ and $T_B^0 \equiv T_B$ since it is constant due to its assumed infinitely large heat capacity $C_B = \infty$.

Π is a parameter which depends on the material and the geometry of the thermal coupling like cross section, length, number and velocity of the energy carriers or their interaction rate.
n depends on the type of coupling [51][52][53]:

$$\begin{aligned} n &= 2 \quad \text{for an electron-electron coupling} \\ n &= 4 \quad \text{for a phonon-phonon coupling} \\ n &= 5 \quad \text{for an electron-phonon coupling} \end{aligned}$$

The thermal coupling G is used in the differential equations with the definition:

$$P =: G \cdot (T_1 - T_2) \tag{5.4}$$

Combining equation (5.3) and equation (5.4), the dependence of the thermal coupling on the two temperatures is for:

$$\begin{aligned} n = 2: &\quad G = \Pi \cdot (T_1 + T_2) \\ n = 4: &\quad G = \Pi \cdot (T_1^3 + T_1^2 T_2 + T_1 T_2^2 + T_2^3) \\ n = 5: &\quad G = \Pi \cdot (T_1^4 + T_1^3 T_2 + T_1^2 T_2^2 + T_1 T_2^3 + T_2^4) \end{aligned}$$

If, for example, in a first case $T_1 = T_2 = 15\,\text{mK}$, and in a second case T_2 is lowered to $10\,\text{mK}$ ($T_1 = 15\,\text{mK}$), the direct thermal coupling G will change as follows:

$$\begin{aligned} n = 2: &\quad \frac{G_{15\,\text{mK}}}{G_{10\,\text{mK}}} \approx 1.2 \\ n = 4: &\quad \frac{G_{15\,\text{mK}}}{G_{10\,\text{mK}}} \approx 1.7 \\ n = 5: &\quad \frac{G_{15\,\text{mK}}}{G_{10\,\text{mK}}} \approx 1.9 \end{aligned}$$

Heat capacities and thermal couplings depend on the temperatures. In the following, the consequences of this for the two time dependent temperatures T_T and T_A will be discussed.

The temperature rise of the absorber is inversely proportional to its heat capacity and therefore connected to its temperature as:

$$\Delta T_A \sim \frac{1}{C_A} \sim \frac{1}{T_A^3}$$

Thus for a lower absorber temperature its temperature rise is increased.

The influence of the changed thermal couplings due to the heater onto the two temperatures is more complex. Therefore a simplified model will be analyzed.

5.3 Equations for the Temperatures

Assuming a physical system consisting of only an infinitely large heat bath, a thermometer with its intrinsic heat capacity only, and a thermal coupling in between, the differential equation can be written in two ways:

$$C_T \cdot \dot{T}_T + G_{BT} \cdot (T_T - T_B) = P_T + P_{Htr}$$
$$\Leftrightarrow \quad C_T \cdot \dot{T}_T + \Pi \cdot (T_T^n - T_B^n) = P_T + P_{Htr}$$

For $t \leq 0$:
$$P_{Htr} = \Pi \cdot ((T_T^0)^n - T_B^n)$$

T_T^0 is the thermometer temperature in the steady state. This simplifies the differential equation:

$$C_T \cdot \dot{T}_T + \Pi \cdot (T_T^n - (T_T^0)^n) = P_T$$

The differential equation, and therefore its solution, is independent of the bath temperature. The increase of the thermal coupling caused by the heater is compensated by itself due to the increase of the power flow caused by the heater. Generally one can say:

> **The temperature development of two coupled systems out of steady state is independent of the lower temperature.**

Applying this to the previous temperature system of figure 5.2 shows that only couplings are affected, where the higher temperature is changed due to the bath temperature change. In case of the CRESST-II light detectors it has to be taken into account that the couplings are usually series connections of different coupling types. Therefore all three couplings are affected differently:

- The thermometer bath coupling G_{BT} is dominated by an electron-electron coupling. Since the thermometer temperature does not change for a bath temperature change, the time constant of this coupling is not changed.

- The absorber thermometer coupling G_{AT} is mainly a phonon-phonon coupling in series with an electron-phonon coupling. The electron-phonon coupling in the thermometer is not affected, since the thermometer temperature is constant. Since the thermometer phonon temperature changes, due to the bath temperature change, also the phonon-phonon coupling between absorber and thermometer changes.

- The absorber bath coupling G_{AB} is dominated by a series of a phonon-phonon coupling and an electron-phonon coupling, too. For both couplings the higher temperature is changed. Therefore both time constants of the coupling are changed.

However, in the CRESST-II experiment **these effects are small**, since the measured signals of the light detectors are dominated by the thermometer-bath coupling and the thermometer's heat capacity. These parameters are not influenced by the thermometer heater.
In any case a constant bath temperature, as used in CRESST-II at the moment, avoids this influence and is introduced for this reason.

5.3.3 Bias Current Self-Heating

As seen in equation (4.3), the mean measured value of the signal is proportional to the bias current through the thermometer I_T. Therefore a large bias current is preferred and for this reason bias current effects have to be taken into account. In this section the **self-heating** of the bias current will be described.

The differential equations describe the temperature change of the thermometer which is used as detector. To read out this thermometer temperature a bias current I_T is flowing through the thermometer. This bias current dissipates a power $\boldsymbol{P_{Bias}}$ in the thermometer of resistance R_T:

$$P_{Bias}(t) = R_T(t) \cdot I_T(t)^2 \tag{5.5}$$

In the steady state the power input is constant:

$$P^0_{Bias} = R^0_T \cdot I^{0\,2}_T \tag{5.6}$$

R^0_T is the steady state resistance of the thermometer, i.e. the operating point where it is stabilized, and I^0_T is the corresponding current through the thermometer.
After an energy deposition the thermometer warms up so that the system leaves the steady state. The resistance $R_T(t)$ and the bias current $I_T(t)$ change. The change of the power dissipation $P^{\Delta T}_{Bias}(t) := P_{Bias}(t) - P^0_{Bias}$ will be derived next.

For small temperature changes the thermometer resistance $R_T(t)$ from (5.5) can be written as:

$$R_T(t) = R^0_T + m^* \cdot (T_T(t) - T^0_T) =: R^0_T + m^* \cdot \Delta T_T(t) \tag{5.7}$$

m^* is the transition slope in the R-T-plane.[8]

The time dependent thermometer bias $I_T(t)$ from equation (5.5) is given as, cp. figure 4.3(b):

$$I_T(t) = \frac{I_{Tot}}{1 + \dfrac{R^0_T + m^* \cdot \Delta T_T(t)}{R_S}} \tag{5.8}$$

---

[8] In contrast to equation (4.2) m^* describes the relation between the real, but unknown, thermometer temperature and its resistance. m describes the relation between the thermometer temperature without bias current effects and the measured thermometer resistance.

5.3 Equations for the Temperatures

The total bias current I_{Tot} is the constant sum of the two currents through the thermometer (I_T) and through the shunt (I_S). R_S is the shunt resistance.

With equations (5.7), (5.8), and (5.5) follows for the bias power input:

$$P_{Bias}(t) = \frac{I_{Tot}^2 R_S^2 \cdot (R_T^0 + m^* \cdot \Delta T_T(t))}{(R_S + R_T^0 + m^* \cdot \Delta T_T(t))^2} \tag{5.9}$$

Thus the constant power dissipation in steady state ($\Delta T_T = 0$) of the bias current into the thermometer is given as, cp. equation (5.6):

$$P_{Bias}^0 = \frac{I_{Tot}^2 R_S^2 R_T^0}{(R_S + R_T^0)^2}$$

The time dependent power input as a function of the temperature change is:

$$\begin{aligned}
P_{Bias}^{\Delta T}(t) &= P_{Bias}(t) - P_{Bias}^0 \\
&= \int_0^{\Delta T_T} \frac{\partial (P_{Bias})}{\partial (\Delta T_T)} d(\Delta T_T) \\
&\stackrel{m^* \cdot \Delta T_T \ll R_S}{\approx} \underbrace{I_{Tot}^2 R_S^2 \cdot m^* \cdot \frac{R_S - R_T^0}{(R_S + R_T^0)^3}}_{=: G_{ETF}} \cdot \Delta T_T \\
&= G_{ETF} \cdot \Delta T_T
\end{aligned}$$

For small thermometer resistance changes, $m^* \cdot \Delta T_T \ll R_S$, the dynamic change of the bias heating $P_{Bias}^{\Delta T}$ is proportional to the temperature difference of the thermometer ΔT_T. This effect is called **electro-thermal feedback**. It behaves like an additional thermal coupling. This coupling G_{ETF} is fictitious and depends on $R_S - R_T^0$, it can be positive or negative. From figure 4.7 it can be seen that the amount for self-heating is maximal for $R_T = R_S$. For the change of the bias current heating follows that for $R_T^0 < R_S$ the bias heating increases during a pulse, whereas it decreases for $R_T^0 > R_S$. Consequently the self-heating effect changes the time dependent thermometer temperature.

Summing up all this, the bias current self-heating has two contributions:

$$P_{Bias}(t) = P_{Bias}^0 + P_{Bias}^{\Delta T}(t) = \frac{I_{Tot}^2 R_S^2 R_T^0}{(R_S + R_T^0)^2} + G_{ETF} \cdot \Delta T_T$$

The first part P_{Bias}^0 is constant. It reduces the heater power P_{Htr} needed. The second part $P_{Bias}^{\Delta T}$ is dynamic. It changes the effective coupling of the thermometer to the bath, and in this way the pulse shape and pulse height. I.e. the change of the bias heating during a pulse changes the thermometer temperature (cp. equation (4.1) and appendix (C.3)):

$$\Delta T_T = \frac{r \, E_{LD}}{C_T} + \Delta T_{Bias}(\Delta T_T) \tag{5.10}$$

Introducing the self-heating effect of the bias current into the model modifies the first of the two coupled differential equations:

$$C_T \cdot \dot{T}_T + G_{AT} \cdot (T_T - T_A) + G_{BT} \cdot (T_T - T_B) = P_T + P_{Htr} + P_{Bias} \quad (5.11)$$
$$C_A \cdot \dot{T}_A + G_{AT} \cdot (T_A - T_T) + G_{AB} \cdot (T_A - T_B) = P_A \quad (5.12)$$

5.3.4 Phonon Collectors

Up to now the **phonon collectors** are not included in the model. Phonon collectors are thin superconducting films well below their critical temperature T_c (e.g. $T = 20\,\text{mK} \ll T_c^{Al} = 1\,175\,\text{mK}$) overlapping part of the thermometer film, see picture 4.3(a).

The principle motivation for phonon collectors is that they can increase the fraction of collected energy in the thermometer r, while they do not increase the thermometer's heat capacity C_T, see equation (4.1). In this way the thermometer's temperature rise ΔT_T can be enlarged:

$$\Delta T_T = \frac{r}{C_T} E_{LD}$$

In a phonon collector well below its critical temperature nearly all electrons are combined to Cooper-pairs.[9] Cooper-pairs do not contribute to the heat capacity, i.e. the heat capacity of a superconductor is mainly determined by the phonon system. Since the volume of the phonon collectors is small, the heat capacity is also small and can be neglected compared to the other heat capacities. This is the reason why the heat capacity of the system does not increase significantly using phonon collectors.

An increase of the energy transfer r into the thermometer can be reached with an additional energy transport from the absorber through the phonon collectors into the thermometer: The non-thermal phonons created in the absorber can travel into the phonon collectors. There they can break Cooper-pairs via absorption. These broken Cooper-pairs are called quasi particles. The quasi particles diffuse like a gas through the collector and emit phonons which break further Cooper-pairs. Finally, a total of roughly 50 % [54] of the in the phonon collectors absorbed energy in form of non-thermal phonons is quickly ($< 1\,\mu\text{s}$ [55]) transferred back into the absorber in form of thermal phonons. The rest of the energy is contained in broken Cooper-pairs. These quasi particles can recombine or diffuse slowly ($\approx 1\,\text{ms}$ [40]) into the thermometer film. Via this diffusion the quasi particles transport additional energy into the thermometer. Phonon collectors can thus increase the average fraction r of the energy which is transferred to the thermometer. A signal of such an energy transport due to quasi particle diffusion was measured in a dedicated experiment [40].

The function of phonon collectors can be compared with an energy storage: First they pick up energy via non-thermal phonon absorption. A part of this energy

---
[9]For $T_c^{Al} = 1\,175\,\text{mK}$ and $T = 20\,\text{mK}$ the part of free electrons in a superconductor should theoretically be $\ll e^{-100} \approx 4 \cdot 10^{-44}$.

5.3 Equations for the Temperatures

is quickly given back to the absorber via thermal phonons. The other part is transported into the thermometer via diffusion of quasi particles with the diffusion time constant τ_D.

Thus the implementation of phonon collectors into the model modifies the power inputs into the thermometer and absorber as follows, cp. equation (5.1) and (5.2):

$$P_T(t) = \varepsilon \cdot (1 - \varepsilon_C) \cdot E_{LD} \cdot \frac{1}{\tau_N} e^{\left(-\frac{t}{\tau_N}\right)} \cdot \Theta(t)$$

$$+ d\varepsilon_C \cdot E_{LD} \cdot \frac{1}{\tau_D} e^{\left(-\frac{t}{\tau_D}\right)} \cdot \Theta(t) \quad (5.13)$$

$$P_A(t) = ((1-\varepsilon) + (\varepsilon - d)\,\varepsilon_C) \cdot E_{LD} \cdot \frac{1}{\tau_N} e^{\left(-\frac{t}{\tau_N}\right)} \cdot \Theta(t) \quad (5.14)$$

The life time of the non-thermal phonons τ_N is:

$$\frac{1}{\tau_N} = \frac{1}{\tau_T} + \frac{1}{\tau_A} + \frac{1}{\tau_C}$$

τ_C is the degradation time of the non-thermal phonons caused by the phonon collectors.

ε_C is the fraction of non-thermal phonons which are absorbed in the phonon collectors:

$$\varepsilon_C = \frac{\tau_A \tau_T}{\tau_A \tau_T + \tau_A \tau_C + \tau_T \tau_C}$$

d is the fraction of $\varepsilon_C \cdot E_{LD}$ which is transported via quasi particles into the thermometer.

Phonon collectors change the power inputs. The input into the absorber is reduced, if the probability for diffusion d is larger than the probability for direct absorption in the thermometer ε, cp. equation (5.14). This power is instead transferred into the thermometer with the time constant τ_D, cp. equation (5.13). In this way the power input into the thermometer can be increased without increasing its heat capacity.

> An increase of the energy $r\,E_{LD}$ which warms up the thermometer will be achieved by phonon collectors as long as the probability for quasi particle diffusion d into the thermometer is larger than the probability for direct non-thermal absorption ε.

However, up to now no significant contribution of the thermometer temperature change caused by phonon collectors could be observed in the CRESST-II experiment [28]:

$$d \approx 0$$

Introducing additionally

$$\boxed{\varepsilon^* := \varepsilon \cdot (1 - \varepsilon_C)}$$

simplifies equation (5.13) and (5.14) into:

$$P_T(t) = \varepsilon^* \cdot E_{LD} \cdot \frac{1}{\tau_N} e^{\left(-\frac{t}{\tau_N}\right)} \cdot \Theta(t) \quad (5.15)$$

$$P_A(t) = (1 - \varepsilon^*) \cdot E_{LD} \cdot \frac{1}{\tau_N} e^{\left(-\frac{t}{\tau_N}\right)} \cdot \Theta(t) \quad (5.16)$$

5.3.5 Time Duration of the Energy Deposition

In all previous equations, as in (5.15) and (5.16), the energy deposition in the light absorber is assumed to be delta function like. This is a good approximation as long as the **duration of the energy deposition** is short compared to the **response time of the thermometer** temperature. Whether this is the case will be analyzed first in this section.

For light detectors the **duration of the energy deposition** is dominated by the **scintillation time constant** τ_S **of CaWO$_4$** at ultra-low temperatures ($\approx 10\,\mathrm{mK}$). This decay time τ_S will be determined in section 5.5.5 to be in the range:

$$400\,\mathrm{\mu s} \lesssim \tau_S \lesssim 500\,\mathrm{\mu s}$$

This decay time is consistent with the value determined in [56] of $\approx 400\,\mathrm{\mu s}$ and with two measurements in [57] of $\approx 400\,\mathrm{\mu s}$ and $\approx 600\,\mathrm{\mu s}$, respectively.

On the other hand, the **response time of the thermometer** is dominated by the **non-thermal phonon life time** τ_N for delta like energy depositions in the light absorber, see below. This value is individual for each light detector and will be determined in section 5.5.5 for five different light detectors. Values are found to be in the range of:

$$200\,\mathrm{\mu s} \lesssim \tau_N \lesssim 400\,\mathrm{\mu s}$$

These values are consistent with the values of three light detectors determined in [28] of 80 - 450 μs.

Thus the approximation $\tau_S \ll \tau_N$ is **not** valid. This can also be seen in figure 5.4. The pulse rising faster is caused by an energy deposition directly in the light absorber. In this case the non-thermal phonon life time determines the rise time.

5.3 Equations for the Temperatures

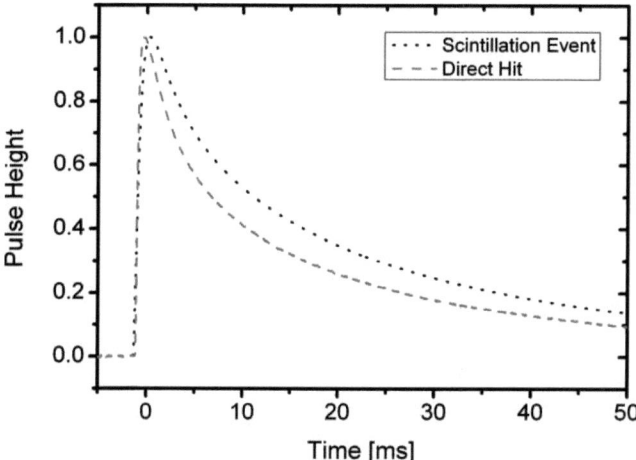

Figure 5.4: In this plot two averaged and normalized pulses of the same light detector are shown. They differ in the energy deposition time duration. In case of the black dotted pulse the energy was deposited in the CaWO$_4$ target crystal, whereas the energy was deposited directly into the light absorber in case of the gray dashed pulse. For the gray dashed pulse the rise time of the pulse is dominated by the non-thermal phonon life time τ_N. On the other hand, the rise time of the black dotted pulse is dominated by the slower scintillation time τ_S of the CaWO$_4$ target crystal at ultra-low temperatures, which limits the energy flow into the light absorber.

The second slower rising pulse is caused by an energy deposition in the scintillating crystal. In this case the decay time of the scintillation process limits the energy flow to the light absorber and in this way the rise time of the thermometer.

In section 4.2 the fraction of absorbed scintillation light is described by q. Therefore, a time dependent light absorption, as it is the case for scintillation light, can be modeled by a time-dependent absorption fraction: $q = q(t)$. For $\tau_S \ll \tau_N$ this would make no difference. However, in practice the situation is opposite:

$$\tau_S \gtrsim \tau_N$$

To include the CaWO$_4$ scintillation duration the power inputs $P_T(t)$ and $P_A(t)$ of the model have to be modified. In general three different cases of energy deposition have to be distinguished:

Single energy deposition: This is the case, for example, for direct gamma absorption in the light absorber. The approximation of a delta like energy deposition is very good. The model does not have to be changed.

Few energy depositions within the pulse duration: This can be the case, for example, for a single photon sensitive light detector absorbing a few photons produced, for example, by a low energetic nuclear recoil in the target crystal. The thermometer response is a superposition of a few single energy depositions. For the total signal modeling knowledge of the arriving times of the photons is necessary. With this time information the response pulses of the thermometer can be determined.

Many energy depositions within the pulse duration: This is usually the case, for example, for gamma absorption in the $CaWO_4$ target crystal. Many photons are produced in the target crystal and are absorbed by the light absorber. The energy deposition can be approximated by a continuous power flow $\sim exp\left(-\frac{t}{\tau_S}\right)$. Thus the power inputs into the thermometer and absorber can be derived using a convolution of the scintillation duration and the non-thermal phonon life time:

$$P_T(t) = \int_{-\infty}^{\infty} \varepsilon^* \cdot E_{LD} \cdot \frac{1}{\tau_S} e^{\left(-\frac{u}{\tau_S}\right)} \cdot \Theta(u) \cdot \frac{1}{\tau_N} e^{\left(-\frac{t-u}{\tau_N}\right)} \cdot \Theta(t-u) \, du$$

$$= \varepsilon^* \cdot E_{LD} \cdot \frac{1}{\tau_S - \tau_N} \left(e^{\left(-\frac{t}{\tau_S}\right)} - e^{\left(-\frac{t}{\tau_N}\right)} \right) \cdot \Theta(t) \qquad (5.17)$$

$$\overset{\tau_S \gg \tau_N}{\approx} \varepsilon^* \cdot E_{LD} \cdot \frac{1}{\tau_S} e^{\left(-\frac{t}{\tau_S}\right)} \cdot \Theta(t) \qquad (5.18)$$

$$P_A(t) = (1-\varepsilon^*) \cdot E_{LD} \cdot \frac{1}{\tau_S - \tau_N} \left(e^{\left(-\frac{t}{\tau_S}\right)} - e^{\left(-\frac{t}{\tau_N}\right)} \right) \cdot \Theta(t) \qquad (5.19)$$

$$\overset{\tau_S \gg \tau_N}{\approx} (1-\varepsilon^*) \cdot E_{LD} \cdot \frac{1}{\tau_S} e^{\left(-\frac{t}{\tau_S}\right)} \cdot \Theta(t) \qquad (5.20)$$

Comparing these two equations, (5.18) and (5.20), with (5.15) and (5.16) shows that in case of $\tau_S \gg \tau_N$ only the time constants of the power flows are exchanged. Depending on the light absorber and the corresponding non-thermal phonon life time, this can be a good approximation.

Usually, the situation is: $\tau_S \gtrsim \tau_N$. Therefore both time constants have to be taken into account for an **exact** modeling of the temperature behavior. In this case the power inputs are given as shown in equations (5.17) and (5.19). Therefore the solution which will be given in section 5.4, has to be extended to the one shown at the end of appendix B, where both time constants are included. Hence there is one more parameter in the solution. However, τ_N can be fixed analyzing direct absorber hits where τ_S does not appear. In this way the number of free parameters for scintillation light events can be reduced by one again. This method is used in section 5.5.5 to determine the parameters relevant to the light detector.

However, for improved clearness the influence of τ_S is not included in the following formulas.

5.3.6 Equations for the Relative Temperatures

As seen in equation (4.3), it is not necessary to know the absolute temperature of the thermometer. It is enough to measure its **relative temperature change** ΔT_T:

$$\Delta T_T := T_T - T_T^0$$
$$\Delta T_A := T_A - T_A^0$$

Introducing these relative temperatures into the two dynamic coupled differential equations (5.11) and (5.12), removes all constant terms:

$$\boxed{C_T \cdot \Delta \dot{T}_T + (G_{AT} + G_{BT}^*) \cdot \Delta T_T - G_{AT} \cdot \Delta T_A = P_T} \quad (5.21)$$
$$\boxed{C_A \cdot \Delta \dot{T}_A + (G_{AT} + G_{AB}) \cdot \Delta T_A - G_{AT} \cdot \Delta T_T = P_A} \quad (5.22)$$

Here $G_{BT}^* := G_{BT} - G_{ETF}$. However, in CRESST-II the electro-thermal feedback is a small effect [28]. Therefore $G_{BT}^* \approx G_{BT}$.

P_T and P_A are given in equation (5.17) and (5.19) for a large number photon absorptions, which is usually the case of interest.

> These two equations are the main equations of the model.
> They describe the temperature behavior
> of the absorber and of the thermometer during a pulse.

5.4 Solution of the Equations

After an energy deposition in the target the temperatures of the target and the light absorber thermometer change. The light detector thermometer response, ΔT_T, can be described by a differential equation. This differential equation is equation (5.21). It is coupled with differential equation (5.22) which describers the temperature behavior of the light absorber, ΔT_A. In the following, the **solution** of these two coupled differential equations will be presented.[10, 11]

---

[10]The ansatz to solve the two equations is presented in appendix B.
[11]For simplified notation in the following the influence of τ_S will not be included. The solution in case of $\tau_S \approx \tau_N$ is given in appendix B.

Solution

The solution of the two differential equations for ΔT_T and ΔT_A is given as:

$$\Delta T_T = \left[A_+ \left(e^{-\frac{t}{\tau_+}} - e^{-\frac{t}{\tau_N}} \right) + A_- \left(e^{-\frac{t}{\tau_-}} - e^{-\frac{t}{\tau_N}} \right) \right] \cdot \Theta(t) \qquad (5.23)$$

$$\Delta T_A = \left[B_+ \left(e^{-\frac{t}{\tau_+}} - e^{-\frac{t}{\tau_N}} \right) + B_- \left(e^{-\frac{t}{\tau_-}} - e^{-\frac{t}{\tau_N}} \right) \right] \cdot \Theta(t) \qquad (5.24)$$

Temperature Parameters

Here are the parameters $A_+, A_-, B_+,$ and B_- given as (cp. appendix B):

$$A_+ = \frac{1}{1 - \frac{\tau_N}{\tau_+}} \cdot \frac{1}{\frac{1}{\tau_-} - \frac{1}{\tau_+}} \cdot \left[\varepsilon^* \cdot \left(\frac{1}{\tau_-} - \frac{G_{AT} + G_{BT}^*}{C_T} \right) + (1-\varepsilon^*) \cdot \frac{G_{AT}}{C_A} \right] \cdot \frac{\mathbf{E_{LD}}}{\mathbf{C_T}}$$

$$=: \frac{1}{1 - \frac{\tau_N}{\tau_+}} \cdot a_+ \cdot \frac{E_{LD}}{C_T} \qquad (5.25)$$

$$A_- = \frac{1}{1 - \frac{\tau_N}{\tau_-}} \cdot \frac{1}{\frac{1}{\tau_+} - \frac{1}{\tau_-}} \cdot \left[\varepsilon^* \cdot \left(\frac{1}{\tau_+} - \frac{G_{AT} + G_{BT}^*}{C_T} \right) + (1-\varepsilon^*) \cdot \frac{G_{AT}}{C_A} \right] \cdot \frac{\mathbf{E_{LD}}}{\mathbf{C_T}}$$

$$=: \frac{1}{1 - \frac{\tau_N}{\tau_-}} \cdot a_- \cdot \frac{E_{LD}}{C_T} \qquad (5.26)$$

$$B_+ = \frac{1}{1 - \frac{\tau_N}{\tau_+}} \cdot \frac{1}{\frac{1}{\tau_-} - \frac{1}{\tau_+}} \cdot \left[(1-\varepsilon^*) \cdot \left(\frac{1}{\tau_-} - \frac{G_{AT} + G_{AB}}{C_A} \right) + \varepsilon^* \cdot \frac{G_{AT}}{C_T} \right] \cdot \frac{\mathbf{E_{LD}}}{\mathbf{C_A}}$$

$$=: \frac{1}{1 - \frac{\tau_N}{\tau_+}} \cdot b_+ \cdot \frac{E_{LD}}{C_A} \qquad (5.27)$$

$$B_- = \frac{1}{1 - \frac{\tau_N}{\tau_-}} \cdot \frac{1}{\frac{1}{\tau_+} - \frac{1}{\tau_-}} \cdot \left[(1-\varepsilon^*) \cdot \left(\frac{1}{\tau_+} - \frac{G_{AT} + G_{AB}}{C_A} \right) + \varepsilon^* \cdot \frac{G_{AT}}{C_T} \right] \cdot \frac{\mathbf{E_{LD}}}{\mathbf{C_A}}$$

$$=: \frac{1}{1 - \frac{\tau_N}{\tau_-}} \cdot b_- \cdot \frac{E_{LD}}{C_A} \qquad (5.28)$$

Time Parameters

The two time constants τ_+ and τ_- are given as (cp. appendix B):

$$\tau_{+,-} = \frac{2}{a \pm \sqrt{a^2 - 4b}} \qquad (5.29)$$

a and b are defined as:

$$a := \frac{G_{AT} + G_{BT}^*}{C_T} + \frac{G_{AT} + G_{AB}}{C_A} \qquad (5.30)$$

$$b := \frac{G_{AT} G_{BT}^* + G_{AT} G_{AB} + G_{BT}^* G_{AB}}{C_T C_A} \qquad (5.31)$$

Here all six parameters of the solution (5.23) and (5.24) are expressed in terms of the heat capacities C_A and C_T, the thermal couplings G_{AT}, G_{AB}, and G_{BT}^*, the deposited energy E_{LD}, the life time of the non-thermal phonons τ_N, and the fraction of non-thermal phonons absorbed in the thermometer ε^*.

5.5 Discussion of the Solution

Out of the two described temperatures the thermometer temperature change ΔT_T is the relevant measured value, cp. equation (4.3). Therefore the focus will be mainly on equation (5.23).

Together with (5.25) and (5.26), equation (5.23) can be written for $t \geq 0$ as:

$$\Delta T_T = \underbrace{\left[a_+ \cdot \frac{e^{-\frac{t}{\tau_+}} - e^{-\frac{t}{\tau_N}}}{1 - \frac{\tau_N}{\tau_+}} + a_- \cdot \frac{e^{-\frac{t}{\tau_-}} - e^{-\frac{t}{\tau_N}}}{1 - \frac{\tau_N}{\tau_-}} \right]}_{\equiv r} \cdot \frac{E_{LD}}{C_T} \quad (5.32)$$

- E_{LD}/C_T is the maximum possible temperature rise of the thermometer in case of 100 % transfer of the absorbed energy into the thermometer: $r = 1$

- r (<1) is the reduction factor of this temperature increase.
  The comparison with equation (4.1) shows[12] that this is the average fraction of the energy absorbed by the light detector, which is transferred into the thermometer film:

$$r = a_+ \cdot \frac{e^{-\frac{t}{\tau_+}} - e^{-\frac{t}{\tau_N}}}{1 - \frac{\tau_N}{\tau_+}} + a_- \cdot \frac{e^{-\frac{t}{\tau_-}} - e^{-\frac{t}{\tau_N}}}{1 - \frac{\tau_N}{\tau_-}}$$

- Both, a_+ and a_-, contain a contribution of non-thermal phonons absorbed directly in the thermometer (ε^*) and a contribution of phonons thermalized in the absorber ($1 - \varepsilon^*$). While a_+ can be **positive and negative**, a_- is always **positive**:

$$a_+ \gtreqless 0$$
$$a_- \geq 0$$

In the following, different special cases for thermal couplings and heat capacities will be discussed which illustrate the solution. They allow a connection of the mathematical parameters and the corresponding physical properties.

---

[12]In ΔT_T of equation (5.32) the current bias effects are included ($\to G_{BT}^*$), while they are not included in ΔT_T^{dep} of equation (4.1) ($\to G_{BT}$).

5.5.1 First Case: $G_{AT} \equiv 0$

In the first case absorber and thermometer are thermally disconnected:

$$\boxed{G_{AT} \equiv 0}$$

In this case the thermometer will be warmed up by $\varepsilon^* \cdot E_{LD}$ and cooled down across G_{BT}^* again. The absorber will be warmed up with $(1-\varepsilon^*) \cdot E_{LD}$ and cooled down across G_{AB}.

The mathematical solution of the time constants simplifies to:

$$\tau_{+,-} = \frac{2}{a \pm \sqrt{a^2 - 4b}} = \frac{C_T}{G_{BT}^*}, \frac{C_A}{G_{AB}}$$

In this special case the correspondence between a time constant and one specific subsystem (absorber or thermometer) is not fixed.[13] The first of the two solutions describes the speed of the temperature change of the thermometer. Assuming that one is the faster one, it will be connected to τ_+. The second solution describes the speed of the temperature change of the absorber. It will be connected to τ_-:

$$\tau_+ = \frac{C_T}{G_{BT}^*} \qquad \tau_- = \frac{C_A}{G_{AB}}$$

With this choice the temperature parameters simplify to:

$$A_+ = \frac{1}{1 - \frac{\tau_N}{\tau_+}} \cdot \frac{\varepsilon^* E_{LD}}{C_T} \stackrel{\tau_N \ll \tau_+}{\approx} \frac{\varepsilon^* E_{LD}}{C_T}$$

$$A_- = 0$$

$$B_+ = 0$$

$$B_- = \frac{1}{1 - \frac{\tau_N}{\tau_-}} \cdot \frac{(1-\varepsilon^*)E_{LD}}{C_A} \stackrel{\tau_N \ll \tau_-}{\approx} \frac{(1-\varepsilon^*)E_{LD}}{C_A}$$

Thus, for $t \geq 0$, the temperature pulses of the thermometer and the absorber are:

$$\Delta T_T = \frac{\varepsilon^* E_{LD}}{C_T} \cdot \underbrace{\frac{e^{-\frac{t}{\tau_+}} - e^{-\frac{t}{\tau_N}}}{1 - \frac{\tau_N}{\tau_+}}}_{<1}$$

$$\Delta T_A = \frac{(1-\varepsilon^*)E_{LD}}{C_A} \cdot \underbrace{\frac{e^{-\frac{t}{\tau_-}} - e^{-\frac{t}{\tau_N}}}{1 - \frac{\tau_N}{\tau_-}}}_{<1}$$

---

[13] $\sqrt{a^2 - 4b} = \sqrt{\left(\frac{G_{BT}^*}{C_T} - \frac{G_{AB}}{C_A}\right)^2} = \left|\frac{G_{BT}^*}{C_T} - \frac{G_{AB}}{C_A}\right|$. It should also be noted that only in this special case, $G_{AT} \equiv 0$, the two time constants, τ_+ and τ_-, can be equal, cp. appendix B.

5.5 Discussion of the Solution

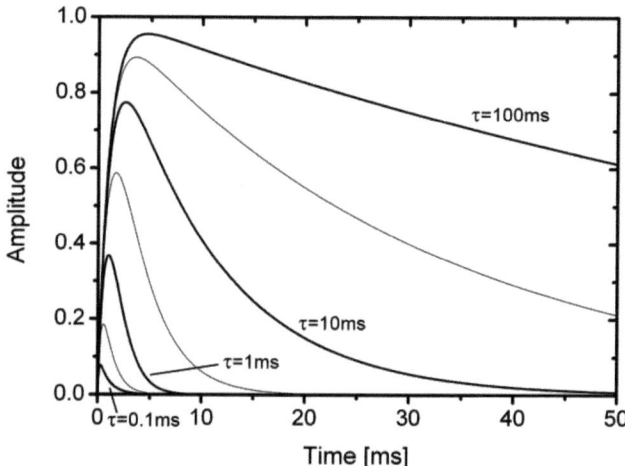

Figure 5.5: In this plot the amplitude of the factor $\dfrac{exp(-t/\tau) - exp(-t/\tau_N)}{1 - \tau_N/\tau}$ is shown for different τ and for $\tau_N = 1$ ms.

The thermometer warms up within the shorter of the two time constants (τ_+ or τ_N) to a temperature which cannot exceed $\frac{\varepsilon^* \cdot E_{LD}}{C_T}$. This maximum possible temperature rise is reduced by the time dependent second factor. This factor approximates to **one** for $\frac{\tau_N}{\tau_+} \to 0$ and to **zero** for $\frac{\tau_N}{\tau_+} \to \infty$, see figure 5.5.[14] Afterwards the temperature relaxes back to the steady state value with the longer of the two time constants (τ_+ or τ_N).
For the absorber τ_+ is exchanged by τ_- and the maximal temperature rise is given by $\frac{(1-\varepsilon^*) \cdot E_{LD}}{C_A}$.

5.5.2 Second Case: $G_{AT} \gg G^*_{BT} = G_{AB}$

The second case describes a thermometer strongly coupled to the absorber. Both, thermometer and absorber have an equal weak coupling to the bath:

$$\boxed{G_{AT} \gg G^*_{BT} = G_{AB}}$$

In this case, after an energy deposition E_{LD}, the energy $\varepsilon^* \cdot E_{LD}$ warms up the thermometer. The energy $(1 - \varepsilon^*) \cdot E_{LD}$ warms up the absorber. Afterwards the temperature difference of thermometer and absorber will be adjusted. Finally, these two temperatures will relax back to the bath temperature within the same time constant.

---

[14] Usually for light detectors the life time of the non-thermal phonons is much shorter than the other two time constants $\tau_N \ll \tau_+, \tau_-$ [28].

In this case, the two time constants are given as:

$$\tau_+ = \frac{2}{a + \sqrt{a^2 - 4b}} \approx \frac{1}{a} = \frac{1}{G_{AT}\left(\frac{1}{C_A} + \frac{1}{C_T}\right)}$$

$$\tau_- = \frac{2}{a - \sqrt{a^2 - 4b}} \approx \frac{a}{b} = \frac{C_A + C_T}{2 \cdot G_{BT}^*}$$

The temperature parameters are:

$$A_+ = \frac{1}{1 - \frac{\tau_N}{\tau_+}} \cdot \left[\frac{\varepsilon^* E_{LD}}{C_T} - \frac{E_{LD}}{C_A + C_T}\right] \stackrel{\tau_N \ll \tau_+}{\approx} \frac{\varepsilon^* E_{LD}}{C_T} - A_-$$

$$A_- = \frac{1}{1 - \frac{\tau_N}{\tau_-}} \cdot \frac{E_{LD}}{C_A + C_T} \stackrel{\tau_N \ll \tau_-}{\approx} \frac{E_{LD}}{C_A + C_T}$$

$$B_+ = \frac{1}{1 - \frac{\tau_N}{\tau_+}} \cdot \left[\frac{(1-\varepsilon^*) E_{LD}}{C_A} - \frac{E_{LD}}{C_A + C_T}\right] \stackrel{\tau_N \ll \tau_+}{\approx} \frac{(1-\varepsilon^*) E_{LD}}{C_A} - B_-$$

$$B_- = \frac{1}{1 - \frac{\tau_N}{\tau_-}} \cdot \frac{E_{LD}}{C_A + C_T} \stackrel{\tau_N \ll \tau_-}{\approx} \frac{E_{LD}}{C_A + C_T}$$

The time dependent temperatures of thermometer and absorber are then given as:

$$\Delta T_T = \left[\frac{\varepsilon^* E_{LD}}{C_T} - \frac{E_{LD}}{C_A + C_T}\right] \cdot \underbrace{\frac{e^{-\frac{t}{\tau_+}} - e^{-\frac{t}{\tau_N}}}{1 - \frac{\tau_N}{\tau_+}}}_{<1} + \frac{E_{LD}}{C_A + C_T} \cdot \underbrace{\frac{e^{-\frac{t}{\tau_-}} - e^{-\frac{t}{\tau_N}}}{1 - \frac{\tau_N}{\tau_-}}}_{<1}$$

$$\Delta T_A = \left[\frac{(1-\varepsilon^*) E_{LD}}{C_A} - \frac{E_{LD}}{C_A + C_T}\right] \cdot \underbrace{\frac{e^{-\frac{t}{\tau_+}} - e^{-\frac{t}{\tau_N}}}{1 - \frac{\tau_N}{\tau_+}}}_{<1} + \frac{E_{LD}}{C_A + C_T} \cdot \underbrace{\frac{e^{-\frac{t}{\tau_-}} - e^{-\frac{t}{\tau_N}}}{1 - \frac{\tau_N}{\tau_-}}}_{<1}$$

The second summands of both temperatures are equal. I.e. for $t \gg \tau_+$ the thermometer and the absorber have the same temperature, since in this case the two first summands are gone.[15] The two first summands describe the differences between the temperature expressed by the second summand and the real temperature. That difference is due to the energy depositions: $\varepsilon^* \cdot E_{LD}$ and $(1-\varepsilon^*) \cdot E_{LD}$.

For $\varepsilon^* > \frac{C_T}{C_A + C_T}$ the first summand of the thermometer is **positive** and the one of the absorber is **negative**. This means that the thermometer is **overheated** compared to the adjusted temperature afterwards. The absorber is instead **undercooled** relative to the absorber's temperature after adjusting to each other. Therefore, the first summand of the absorber is negative. For $\varepsilon^* < \frac{C_T}{C_A + C_T}$ it is the other way round.

---
[15] Assuming $\tau_N \ll \tau_+, \tau_-$, which is usually the case for light detectors.

5.5 Discussion of the Solution

Within the time τ_N, thermometer and absorber are warmed up to different temperatures. After the time τ_+ they have adjusted to each other. After τ_- they have relaxed back to their steady state temperatures, since the energy has been transported to the bath.

5.5.3 Third Case: $C_T \ll C_A$ & $G_{AB} \equiv 0$

In the third case the thermometer is almost ideal, i.e. with only little influence onto the system:
$$\boxed{C_T \ll C_A}$$
Additionally, the absorber is not connected to the bath:
$$\boxed{G_{AB} \equiv 0}$$

In this case the thermometer will be warmed up by $\varepsilon^* \cdot E_{LD}$ within a time of about τ_N.[16] The absorber will be warmed up within the same time by the energy $(1-\varepsilon^*) \cdot E_{LD}$. Afterwards the thermometer temperature changes quickly since its heat capacity is very small. It will change to a value in between the bath and the absorber temperature, since the absorber is coupled to the bath only through the thermometer. Both temperatures will relax back to the initial temperatures with the same time constant.

Mathematically, the two time constants are in this case:
$$\tau_+ = \frac{2}{a + \sqrt{a^2 - 4b}} \approx \frac{1}{a} = \frac{C_T}{G_{AT} + G_{BT}^*}$$
$$\tau_- = \frac{2}{a - \sqrt{a^2 - 4b}} \approx \frac{a}{b} = \left(\frac{1}{G_{AT}} + \frac{1}{G_{BT}^*}\right) C_A$$

The fast time constant τ_+ is the connection of the two independent couplings connected to the heat capacity C_T. It describes the speed of the temperature change of the thermometer due to the two connected thermal couplings.

The slow time constant τ_- is the series connection of the two couplings between the heat capacity C_A and the bath. It describes the speed of the temperature change of the absorber, i.e. it is the absorber relaxation time.

As seen in equation (5.23), the amplitudes A_- and A_+ together with the exponential functions express the temperature change of the thermometer. In this case the first amplitude is given as:
$$A_- = \frac{1}{1 - \frac{\tau_N}{\tau_-}} \cdot \frac{1}{1 + \frac{G_{BT}^*}{G_{AT}}} \cdot \frac{(1-\varepsilon^*)E_{LD}}{C_A}$$
$$\underset{\tau_N \ll \tau_-}{\approx} \frac{1}{1 + \frac{G_{BT}^*}{G_{AT}}} \cdot \frac{(1-\varepsilon^*)E_{LD}}{C_A}$$

---
[16] Assuming the life time of the non-thermal phonons τ_N to be the shortest time constant as it is usually the case for CRESST-II light detectors [28].

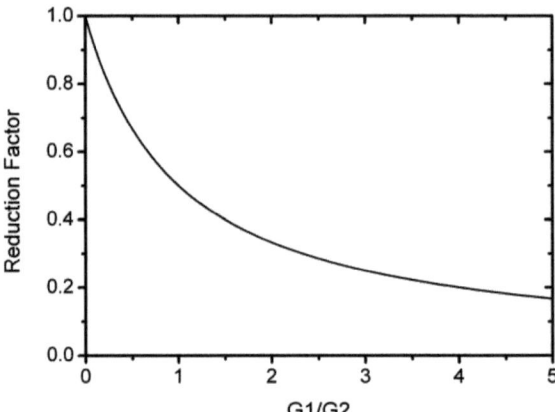

Figure 5.6: In this plot the reduction factor $1/\frac{1}{1+G_1/G_2}$ is shown as a function of G_1 and G_2. For $G_1 \ll G_2$ it is ≈ 1. For $G_1 = G_2$ it is $1/2$ and for $G_1 \gg G_2$ it approaches zero.

The first factor reduces the maximal possible temperature rise together with the time dependent exponential functions as in the cases before. This depends on the time duration of the power input τ_N and the time duration of the power output τ_- of the absorber through the thermometer, cp. figure 5.5.

The second factor describes the reduction of this temperature rise caused by the relation of the two couplings G^*_{BT} and G_{AT}. The stronger the thermometer is coupled to the absorber (G_{AT}) compared to the coupling to the bath (G^*_{BT}), the closer its temperature is to the absorber's temperature. This reduction factor is illustrated in figure 5.6.

The third factor is the maximal possible temperature rise of the absorber caused by the decayed non-thermal phonons $(1 - \varepsilon^*)$.

The second amplitude is given as:

$$A_+ = \frac{1}{1 - \frac{\tau_N}{\tau_+}} \left[\frac{\varepsilon^* E_{LD}}{C_T} - \frac{1}{1 + \frac{G^*_{BT}}{G_{AT}}} \cdot \frac{(1 - \varepsilon^*) E_{LD}}{C_A} \right]$$

$$\overset{\tau_N \ll \tau_+}{\approx} \frac{\varepsilon^* E_{LD}}{C_T} - A_-$$

The first factor is again the time dependent reduction with the exponential functions, cp. figure 5.5.

The first summand in the bracket is the maximal temperature rise of the thermometer caused by the non-thermal phonons absorbed in the thermometer. This temperature rise is reduced by the second summand, which is equal to A_-.[17]

---

[17] Apart from the prefactor of the bracket, which describes the time dependent reduction.

5.5 Discussion of the Solution

A_+ is the **temperature difference** between the temperature given by the thermal position in between absorber and bath (A_-) and the one due to the heating up due to the non-thermal phonon absorptions. Therefore, depending on the fraction ε^* of absorbed non-thermal phonons in the thermometer, A_+ can be positive, zero, or negative:

$$A_+ \begin{cases} < 0 \\ = 0 \\ > 0 \end{cases} \text{ for } \varepsilon^* \begin{cases} < X \\ = X \\ > X \end{cases} \text{ with } X := \frac{1}{1 + \frac{G^*_{BT}}{G_{AT}}} \cdot \frac{C_T}{C_A}$$

In summary the amplitudes can be interpreted as:

- A_- is the **temperature of the thermometer** at times $t \gg \tau_+$.[18] This temperature is given as the absorber's temperature reduced by the thermal position of the thermometer in between of the two couplings to absorber and bath. This reduction factor is shown in figure 5.6. The time duration of this temperature decay is given by the relaxation time of the absorber τ_-.

- A_+ describes the τ_+ **short temperature difference** of the thermometer compared to A_-. This temperature difference is caused by the fraction ε^* of non-thermal phonons, which are absorbed by the thermometer.
  If the thermometer is heated by the non-thermal phonons above A_-, A_+ will describe an **overheating**. A_+ is positive. If the fraction ε^* of absorbed non-thermal phonons is instead not enough to increase ΔT_T over A_-, A_+ will be negative. It then describes an relative **undercooling** of the thermometer.

5.5.4 Summary

Before the parameters and physical values of a real light detector will be determined in the last section of this chapter, previous fact-findings will be recapitulated. Their understanding enables the interpretation and understanding of the behavior of a CRESST-II light detector:

▶ The two coupled differential equations (5.21) and (5.22) describe the temperature behavior of the light detector thermometer and absorber after an energy deposition in the light absorber.

▶ These two time dependent temperatures are parameterized in two equations, (5.23) and (5.24). The connection of these parameters to the physical values are given in (5.25)-(5.31).

▶ The parameters A_- and B_- represent the temperature rises of thermometer and absorber, respectively, after the dynamic adjusting within the time constant τ_+. Together with the exponential functions they describe the

---

[18]This assumes an even shorter non-thermal phonon life time τ_N which is usually given for light detectors in CRESST-II: $\tau_N \ll \tau_+ \ll \tau_-$ [28].

time dependent development of these two temperatures. In all cases, these two parameters are positive:

$$A_-, B_- > 0$$

▶ The absorption and the decay of the non-thermal phonons heat up the thermometer and the absorber. The temperature differences of these temperature rises to A_- and B_- are given by A_+ and B_+, respectively. If the absorber is overheated compared to the dynamic state temperature, the thermometer will be undercooled relative to the thermometer's dynamic state temperature, and vice versa. Therefore

$$sign(A_+) = -sign(B_+)$$

▶ Across the thermal couplings to the heat bath, absorber and thermometer relax back to their initial temperatures. The duration of this relaxation is described by the slower time constant τ_-.

5.5.5 Fourth Case: Real Light Detector

In previous cases, simplified solutions have been analyzed. In this fourth case a real light detector, e.g. the one named Steven, and its physical properties will be specified. In this way the physical values influencing the temperature behavior of a light detector after an energy deposition in the detector target can be determined.

Including the scintillation decay time of the target crystal, **the temperature change depends on nine physical properties**, cp. section 5.4:

$$C_T,\ C_A,\ G_{BT}^*,\ G_{AB},\ G_{AT},\ \tau_N,\ \tau_S,\ \varepsilon^*,\ \text{and } E_{LD}$$

Six of these values can be fixed by fitting the model's solution to two different types of measured thermometer pulses. On the one hand, these are pulses caused by gamma absorptions in the target crystal and, on the other hand, by direct energy depositions in the light absorber. Finally, **three** values have to be derived with the help of further information. For this reason the theoretically expected values for the two heat capacities and the derived absorbed energy are used:

- The temperature dependent heat capacity of the thermometer is determined in appendix A:

$$\boxed{C_T = X \cdot 17.86\,\frac{\text{eV}}{\text{mK}} \cdot \frac{T_T}{\text{mK}}} \quad (1 \leq X \leq 2.43) \quad (5.33)$$

X is in between 1 and 2.43 depending on the operating point in the transition. The thermometer temperature is in millikelvin and can be taken from the operating point.

5.5 Discussion of the Solution

- The derived absorber heat capacity C_A is also estimated in appendix A:

$$\boxed{C_A = 1.23 \, \frac{\text{eV}}{\text{mK}} \cdot \left(\frac{T_A}{\text{mK}}\right)^3} \qquad (5.34)$$

  The temperature is in millikelvin and is determined by the temperatures T_T and T_B, and the two couplings G_{AT} and G_{AB}.

- As third value, E_{LD} is derived. To get this information an absolute energy calibration of the light absorber is done [58]. This provides the fraction

$$\boxed{p\,q = \frac{E_{LD}}{E_{dep}} = 1.88\,\%}$$

  of the energy deposited in the target crystal (E_{dep}) which is absorbed by the light absorber (E_{LD}). $p \cdot q$ depends on the individual detector module, see section 4.2, and is derived in this case for the detector module "Rita-Steven".[19] In this calibration measurement the thermometer changes caused, on the one hand, by well-known energy depositions directly in the light absorber (E_{LD}) and, on the other hand, by well-known energy depositions in the target crystal (E_{dep}), have been compared. In general, the knowledge of $p \cdot q$ enables the knowledge of E_{LD} for a known energy deposition in the target crystal.[20]

With the help of these three values (C_T, C_A, and E_{LD}) the further six values can be determined by fitting averaged measured temperature pulses of the thermometer:

- To determine τ_N, averaged normalized pulses produced by similar energy depositions **directly** into the light absorbers[21] are analyzed. These events, and therefore the solution, are independent of the scintillation time of the target crystal τ_S due to the direct energy deposition into the light absorber. Fitting formula (5.23) to this averaged pulse results in the values listed in table 5.1 on the left hand side. The four light detectors (LD) named Hans, Q, Steven, and X consist of SOS-wafers, while LD Ulrich consists of pure silicon.[22] It can be seen that the life time of the non-thermal phonons is in the range:

$$\boxed{200\,\mu\text{s} \lesssim \tau_N \lesssim 400\,\mu\text{s}}$$

  These values are consistent with the ones derived in [28] of 80 - 450 μs for the light detectors named SOS21, SOS23, and SOS30. It should be noted that the non-thermal phonon life time τ_N is light detector dependent.

---

[19] The phonon detector is named Rita and the light detector Steven.

[20] In the model of this chapter, it is assumed that the total amount of absorbed energy in form of non-thermal phonons heats up absorber or thermometer, cp. footnote 5 of this chapter. In reality, part of the non-thermal phonons is lost due to warming up of the holding clamps and the cooling structure of the light detector.

[21] Pulses taken from [58].

[22] SOS $\triangleq$ silicon on sapphire; i.e. a 460 μm thick sapphire wafer covered on one side by a 1 μm thick silicon light absorption layer.

	Direct Hits				Scintillation Events			
LD	A [%]	τ_+ [ms]	τ_- [ms]	τ_N [ms]	A [%]	τ_+ [ms]	τ_- [ms]	τ_S [ms]
Hans	85	6.59	19.54	0.28	88	7.88	26.62	0.52
Q	81	5.60	30.00	0.29	78	8.69	35.62	0.38
Steven	56	4.58	30.20	0.18	49	6.15	34.05	0.38
X	98	4.13	27.11	0.40	98	4.52	28.37	0.48
Ulrich	54	4.68	20.96	0.41	53	4.20	19.70	0.11

Table 5.1: In this table resulting parameters for five light detector (LD) fits are shown. For the fitting of averaged templates of **direct hit** pulses formula (5.23) is used. For **scintillation event** pulses formula (B.18) is used. $A := A_+/(A_+ + A_-)$. The non-thermal phonon life time τ_N for Ulrich is naturally longer by about a factor of two due to the reduced velocity of sound in silicon compared to sapphire. The decay time of the target crystal τ_S from Ulrich is shorter since it was operated with a $ZnWO_4$ crystal which shows a faster scintillation time. All other light detectors were paired with $CaWO_4$ crystals.

From the first four analyzed light detectors in table 5.1 one can see that the non-thermal phonon life time is correlated with the amount of overheating A:

$$A := \frac{A_+}{A_+ + A_-}$$

The longer the non-thermal phonon life time is, the more of these phonons are absorbed in the thermometer. The fraction of absorbed non-thermal phonons in the thermometer ε^* increases, and in this way the overheating A_+ and A, respectively. In contrast to these four light detectors, Ulrich consists of a pure silicon wafer, where the velocity of sound is about a factor two smaller compared to sapphire. Since non-thermal phonons decay mainly at surface reflections [39], their life time is enlarged by the smaller velocity of sound by this factor of about two. For this reason, Ulrich can be compared with Steven. This is consistent with their two determined values of A which are similar.

- In a second step, averaged, normalized pulses produced by scintillation events[23] are fitted by equation (B.18). The averaged and the fitted pulse from Steven can be seen in figure 5.7. In this fit τ_N is fixed and taken from the fit above. The fit results of the parameters can be seen in table 5.1 on the right hand side. The scintillation time of $CaWO_4$ at ultra-low temperatures ($T \approx 10\,\text{mK}$) is determined to be in the range:

$$\boxed{400\,\mu\text{s} \lesssim \tau_S \lesssim 500\,\mu\text{s}}$$

This value is consistent with the values determined in [28] of $400\,\mu\text{s}$ and in [57] of $400\,\mu\text{s}$ and $600\,\mu\text{s}$.

---

[23]Pulses taken from [58].

5.5 Discussion of the Solution

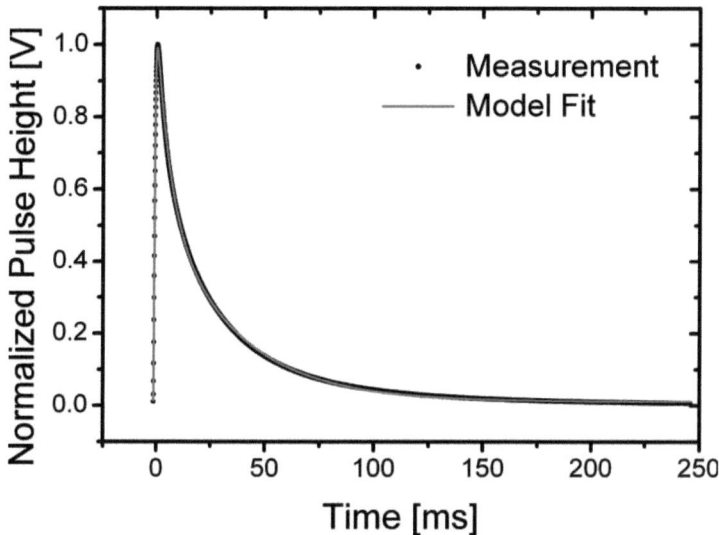

Figure 5.7: In this plot the averaged and normalized measured pulses of Steven can be seen in black. The fitted model pulse of equation B.18 from appendix B with five free parameters is shown in gray.

Apart from that, it can be seen that the decay time of the scintillation light of the target crystal from Ulrich is about:

$$\tau_S \approx 110\,\mu s$$

This is caused by the difference of the scintillating target material. In this case it is **ZnWO$_4$** instead of CaWO$_4$ as it was the case of the four other detectors. For this material the decay time at ultra-low temperatures is naturally shorter.

Above, the five values C_T, C_A, E_{LD}, τ_N, and τ_S have been determined directly. The remaining four values G_{BT}^*, G_{AB}, G_{AT}, and ε^* have to be derived from the four fitted values A_+, A_-, τ_+, and τ_-. Therefore, equations (5.25), (5.26), and (5.29) - (5.31) can be used.

To apply these equations, A_+ and A_- have to be converted first from their measured voltage [V] value into a temperature difference [mK]. This can be done, as explained in appendix C.3, with the help of the transition curve[24] and the knowledge of the operating point.[25]

---

[24] Taken from [59].
[25] The maximum temperature rise of the light detector Steven caused by a gamma absorption of 122 keV in the target crystal is derived to be 294 µK. It should be noted that this results in: $A_+ = 166\,\mu K$ and $A_- = 173\,\mu K$.

X	ε^* [%]	$G_{AT}\left[\frac{\text{pW}}{\text{K}}\right]$	$G_{AB}\left[\frac{\text{pW}}{\text{K}}\right]$	$G_{BT}^*\left[\frac{\text{pW}}{\text{K}}\right]$	T_A [mK]
1.0	3.7	0.7	6.9	6.1	10.9
1.5	5.6	1.1	6.9	9.1	11.1
2.0	7.5	1.6	7.0	12.0	11.3
2.43	9.1	2.0	6.9	14.5	11.4

Table 5.2: In this table the physical values ε^*, G_{AT}, G_{AB}, and G_{BT}^* are listed. These values are derived from the fitted parameters A_+, A_-, τ_+, and τ_-, which result from the fit of formula (B.18) to an averaged, normalized pulse produced by scintillation events. Due to the fact that X is only known to be in between 1 and 2.43, the influence of the choice is tested. For the determination of these physical values, in addition it is used that the thermometer is stabilized at $T_T \approx 14.7$ mK and the bath temperature is $T_B \approx 10.5$ mK. It can be seen that less than 10 % of the non-thermal phonons are absorbed by the thermometer. The coupling of the absorber to the thermometer is the weakest. Thermometer and absorber are thermally decoupled due to the weak electron-phonon coupling at ultra-low temperatures (T^5). This is reflected by the absorber temperature, which is much closer to the bath temperature than to the thermometer temperature.

In case of light detector Steven, solving the previous system of equations results in the values listed in table 5.2. Due to the lack of knowledge on the parameter X which influences the thermometer heat capacity C_T, cp. equation (5.33), and which is connected to the operating point in the transition, the values are derived for different X. In addition, it should be noted that the absorber temperature T_A, which is listed in the table, as well,[26] is between $T_B = 10.5$ mK and $T_T = 14.7$ mK depending on the two couplings G_{AT} and G_{AB}.

It should also be noted that the electron contribution of a coupling depends on the experiment and is not a constant of matter. The thermal coupling of electrons, at these ultra-low temperatures, is dominated by the mean free path length of the electrons which is strongly connected to the material purity. For this reason differences between the electron couplings of different experiments can be expected.

In the following these four values will be discussed and compared with the values measured at other experiments:

- As can be seen, the fraction of non-thermal phonons which is absorbed by the thermometer is, in case of light detector Steven, in the range:

$$3.7\% \lesssim \varepsilon^* \lesssim 9.1\%$$

This is consistent with the value of $\varepsilon^* \approx 5\%$ derived for a light detector in [28].

---

[26] The absorber and the thermometer temperature have to be known, since they influence the heat capacities, see equations (5.33) and (5.34).

5.5 Discussion of the Solution

A similar value can roughly be estimated in a simple way, too: Assuming a speed of sound of 5 km/s for sapphire [39], taking twice the thickness of the absorber (0.92 mm) as path length between two surface hits, then 5.4 % of the non-thermal phonons hit the thermometer once in their life time τ_N. Assuming that all non-thermal phonons which hit the thermometer are transmitted into it and are, due to the strong electron-phonon coupling of non-thermal phonons, absorbed, results in $\varepsilon^* \approx 5\,\%$.

- The thermal coupling between thermometer and absorber is derived to be in the range:

$$\boxed{0.7\,\frac{\mathrm{pW}}{\mathrm{K}} \lesssim G_{AT} \lesssim 2.0\,\frac{\mathrm{pW}}{\mathrm{K}}}$$

This coupling is significantly smaller than the other couplings.
This value can be compared with another measurement [40]. In that measurement the electron-phonon coupling for tungsten at ultra-low temperatures has been derived. This electron-phonon coupling G^{ep} dominates G_{AT}, therefore $G_{AT} \approx G^{ep}$. Extrapolating the measured value with the expected temperature dependence to the thermometer temperature T_T and using the light detector thermometer volume results in a comparable value: $G^{ep} = 4.0\,\mathrm{pW/K}$.

- The coupling of the absorber to the bath is determined to be:

$$\boxed{G_{AB} \approx 6.9 \pm 0.1\,\frac{\mathrm{pW}}{\mathrm{K}}}$$

This value can be compared to a previously measured one. As before, the value determined in [39] is extrapolated with the assumed temperature dependence to the absorber temperature. As interaction volume the volume of the gold pad is taken. I.e. it is assumed that this pad dominates the thermal coupling between absorber and bath G_{AB}. In this way, the couplings through the clamps and the gold bond wire are neglected. The derived value in [39] is: $G_{AB} = 0.2\,\mathrm{pW/K}$. In the same work a theoretical value derived from Pippard's free-electron model is given which results in: $G_{AB} = 0.7\,\mathrm{pW/K}$. Both values are significantly smaller than the result obtained here. The reason therefor can be, as mentioned, that the coupling of the holding clamps was not included and the volume of the gold bond wire was ignored.

- The determined values for the coupling of the thermometer to the bath G^*_{BT} are in the range of:

$$\boxed{G^*_{BT} \approx 10 \pm 5\,\frac{\mathrm{pW}}{\mathrm{K}}}$$

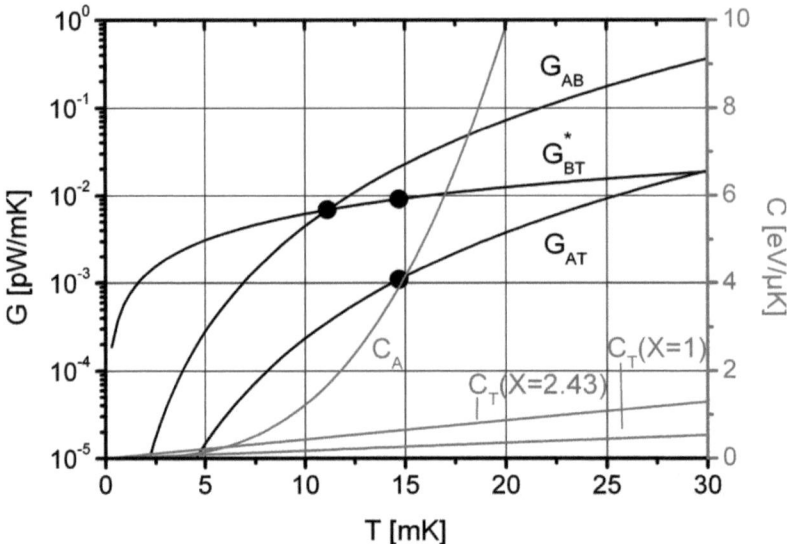

Figure 5.8: In this plot the three determined values of the thermal couplings are shown as dots for $X = 1.5$. Their expected temperature dependence can also be seen (note the logarithmic scale). Additionally, the absorber and thermometer heat capacities are shown. The thermometer's heat capacity is in between the value at the top of the transition ($X = 1$) and the value for being in the total superconducting state ($X = 2.43$).

This value can be compared with a theoretical value derived with the Wiedemann–Franz law: Assuming a resistivity of $2.44 \cdot 10^{-8}\,\Omega\mathrm{m}$ and a residual resistance ratio RRR of 2.5 (which has been measured for another thin gold strip in the frame of this work) results in:[27] $G^*_{BT} \approx G_{BT} \approx 40\,\mathrm{pW/K}$.

To summarize above results of the thermal couplings and the heat capacities, these values are plotted in figure 5.8. The three derived values for the heat capacities are shown as points, assuming $X = 1.5$, whereas their expected temperature dependencies, see section 5.3.2, are shown as lines. The strong temperature dependencies can be seen on the logarithmic scale. Additionally, the two temperature dependent heat capacities are plotted. For the thermometer heat capacity, C_T, the upper ($X = 2.43$) and the lower ($X = 1$) limits are shown.

After the determination of the different physical values (C_T, C_A, G^*_{BT}, G_{AB}, G_{AT}, τ_N, τ_S, ε^*, and E_{LD}) their influence on the solution parameters of the thermometer temperature (τ_+, τ_-, A_+, and A_-) after an energy deposition can be described. In this way, it can be seen that for $\tau_N, \tau_S \ll \tau_+ \ll \tau_-$ the following parameters

---

[27] The influence of the electro thermal feedback can be neglected [28].

5.5 Discussion of the Solution

can be approximated as:

$$\tau_+ \approx \frac{C_T}{G_{BT}^*}$$
$$\tau_- \approx \frac{C_A}{G_{AB}}$$
$$A_+ \approx \frac{\varepsilon^* E_{LD}}{C_T} - A_-$$
$$A_- \approx \frac{G_{AT}}{G_{BT}^*} \cdot \frac{(1-\varepsilon) E_{LD}}{C_A}$$

Due to the weak thermal coupling G_{AT}, absorber and thermometer are strongly connected to one of the two time constants τ_+ and τ_-. The two time constants behave as in the first case, discussed in section 5.5.1. Since the thermometer heat capacity is smaller the thermometer is associated with τ_+.

A_- is dominated by the heating up of the absorber $((1 - \varepsilon^*) E_{LD}/C_A)$ and the thermal position of the thermometer in between absorber and bath, cp. the third case for small G_{AT}/G_{BT}^*. A_+ is the overheating above A_- caused by $\varepsilon^* E_{LD}/C_T$.

Chapter 6

Light Detector Optimization

In chapter 4 and 5 the physical properties which influence the widths of the bands in the light-phonon-plane have been analyzed. This chapter picks up some of these properties with the goal of improving the energy resolution of the light channel and, in this way, the width of the bands.

6.1 Experiments

6.1.1 Black Silicon Light Absorber

As seen in the sections 4.1 and 4.2.2 the fraction q of the created photons which is absorbed by the light absorber depends on many parameters, for example the self-absorption of the crystal, the reflection probability of the surrounding housing, or the absorption probability of the light absorber. These parameters determine the energy transport in form of photons starting from the light production in the target crystal and ending with the light absorption. This section will focus on increasing the **light absorption probability of the light absorber** to raise the fraction q.

For this, a technique is used which has been developed for improving the efficiency of photovoltaic cells [60]. There silicon wafers are used as light absorber. Treating the surface of the silicon increases the light absorption probability by about a factor of two to above 95 % for a wide range of wavelengths ($\approx 200\,\mathrm{nm} - 800\,\mathrm{nm}$). Even larger probabilities and a broader wavelength range can be reached by optimizing the treatment onto the individual silicon material.
The improvement is based on an etching mechanism which produces nano holes in the silicon surface, their diameter is much smaller than the wavelength of the absorbed light. In this way a continuous change of the index of refraction from air or vacuum to silicon is introduced. The sharp change of the index of refraction at the surface, and the reflection probability connected with it is washed out.

In figure 6.1, different light absorbers can be seen. The standard SOS (silicon on sapphire) CRESST-II light absorbers are shown on the left hand side. In the top left picture the silicon absorption layer is on the top, whereas in the bottom

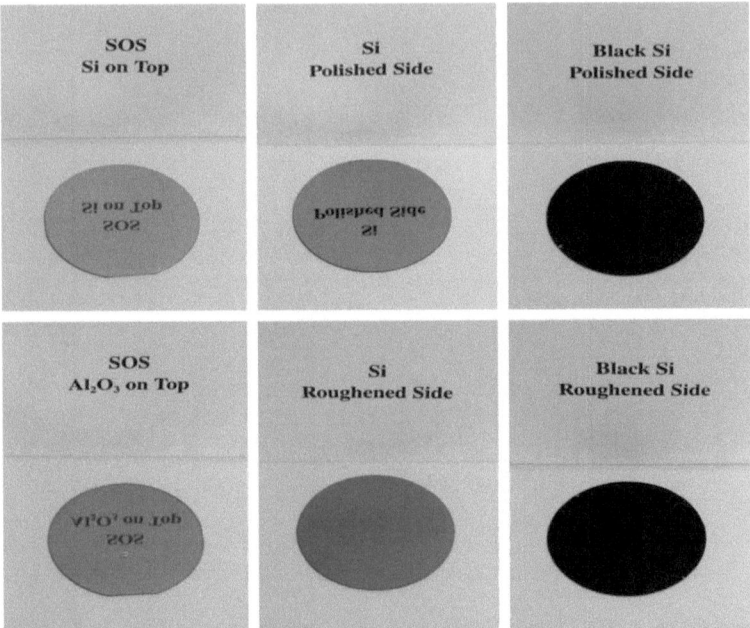

Figure 6.1: Three light absorber, each from both sides. On the left hand side, an SOS (silicon on sapphire) wafer is shown. This standard light absorber is mainly used in CRESST-II using the sapphire (Al_2O_3) side facing the scintillating crystal due to the larger absorption probability [38]. In the middle picture an untreated silicon absorber is shown. In the top picture, the polished side is on the top. The back side, etched by the producer to obtain a roughened surface, can be seen in the bottom picture. Both sides of the silicon wafer are shown after a special etching treatment on the right hand side. This etching provides a continuous change of the index of refraction at the surface which suppresses the reflection probability to below 4 %. By optimizing the treatment process on the individual silicon wafer, even larger absorption probabilities can be reached [60]. The legend from the back plane reflected in the detector indicates the reflectivity, too.

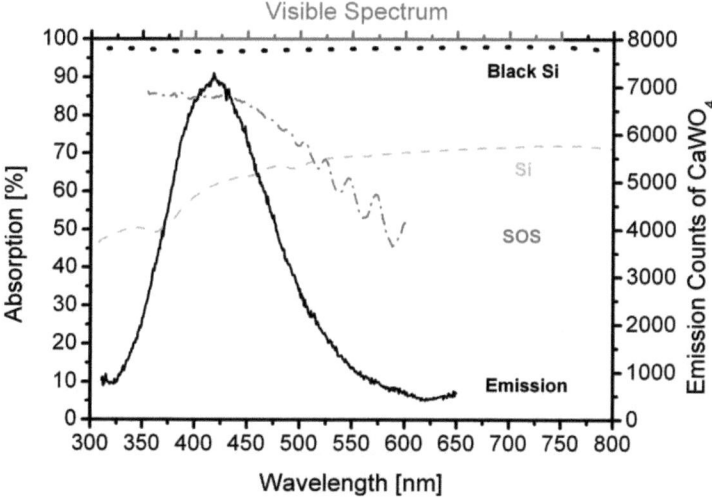

Figure 6.2: In this figure the absorption probabilities of a SOS wafers (sapphire on the top) [38], an untreated silicon [61], and a black etched silicon [61] can be seen. The treatment increases the absorption probability to more than 96 %. Compared to standard light absorbers of CRESST-II (SOS), this is an improvement of about 10-20 % in the relevant energy range of the $CaWO_4$ emission spectrum shown as well. For comparison, the energy range of the visible spectrum is marked on the top axis.

left picture it is the opposite sapphire side.[1] The reflectivity of the absorber can, apart from the color, also be seen from the mirrored writing from the back plane. An untreated silicon wafer can be seen in the middle pictures. On the top middle, the polished side of the silicon wafer is shown. In the bottom middle, the roughened side (obtained via etching by the producer) is placed on the top. The mirrored writing cannot be seen anymore due to diffuse reflection. However, the total reflectivity of these two sides of the silicon wafer is very similar.

On the right hand side, theses two silicon wafers are shown after the etching treatment described in [60]. It can be seen that absorption has been increased impressively.

To quantify these observations the determined absorption probabilities of these three wafer types are compared in figure 6.2. Depending on the wavelength the absorption probabilities of the SOS-wafer (sapphire on top) [38], the untreated silicon (Si) [61], and the treated silicon (black Si) [61] are plotted on the left hand axis.[2] Additionally, on the right hand axis the emission spectrum of the $CaWO_4$ target crystal [36] is shown, which has to be absorbed. To compare the pictures

---
[1] Out of these six wafers, a thermometer structure is only placed on the bottom left absorber.
[2] The difference of absorption probability between the polished and etched sides is negligible.

of figure 6.1, on the top axis the wavelength range of the visible light is shown.
A dramatic improvement of the light absorption probability by treating the silicon wafer was observed (Si to Black Si). The absorption probability of Black Si is above 96.7 % in the wavelength range shown. Over the whole relevant emission spectrum of $CaWO_4$ it is thus superior to the standard SOS-wafer. In addition, for emission spectra with larger wavelengths, as it is the case, for example, for $ZnWO_4$ (425 nm - 575 nm) or $CaMoO_4$ (450 nm - 600 nm) [62], it provides a constant high absorption probability.

> **In the wavelength range relevant for scintillators, black treated silicon absorbers show an excellent performance.**

6.1 Experiments

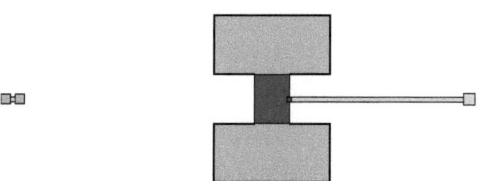

Figure 6.3: In this technical drawing the design of a thermometer structure with a separated heater is shown. In figure 4.3(a) a picture of the standard design can be seen. The dark centered part is the tungsten thermometer. On the upper and the lower end the aluminum phonon collectors are placed. The gold strip, connected on the right hand side of the thermometer, is used for cooling the thermometer. The separated heater, on the left hand side, is placed at a distance of 2 mm to the thermometer. It consist of a 40 µm × 40 µm × 30 nm gold heater and the two aluminum bond bads (80 µm × 80 µm × 1 000 nm) on the right and left hand side. An energy deposition in the heater can be transported through the absorber into the thermometer fast enough to control it thermally.

6.1.2 Separated Heater

Thermometer and heater of a CRESST-II light detector are electrically connected to each other, see figure 4.3(a). The advantage of this strong electron coupling at ultra-low temperatures, cp. section 5.3.2, is the easy thermal stabilization of the thermometer in its operating point. To stabilize a thermometer in the operating point **heater pulses** warm up the light detector. The resulting temperature rise of the thermometer is detected. From this thermometer response the operating point in the transition can be determined. This information is used to control the heater current offset to **adjust the thermometer** to the set operating point.

One drawback of the electric connection between heater and thermometer is that **common mode noise** can appear, as seen in section 4.3.6. External interferences can be picked up between the heater and the bias wires. These interferences are introduced directly into the SQUID read out and, in addition, warm up the thermometer structure. In this way the electric connection between heater and thermometer makes the detector susceptible to electro-magnetic interferences, cp. section 4.3.6.

To avoid this drawback, a new design of the thermometer structure has been tested. In this new structure **thermometer and heater are separated** electrically, as shown in figure 6.3.
As in the previous design, the heater has to fulfill two tasks:

- First, the heater has to warm up the thermometer into its operating point, cp. section 5.3.2.

- Second, via heater pulses the operating point of the thermometer has to be controlled.

In addition, in the new design the thermometer is still coupled via an isolated gold strip to the heat bath. This electron coupling is able to transport enough heat out of the thermometer, so that the read out bias I_T is not limited by its self heating effect.

In order not to absorb many non-thermal phonons after an energy deposition in the light absorber, the area of the heater is chosen to be as small as feasible.

After implementing the new design, the two tasks mentioned before have been tested:

It was confirmed that it is possible to warm up a thermometer through an absorber substrate used in CRESST-II.

On the other hand, absorber and thermometer are only weakly coupled to each other due to the electron-phonon coupling, see section 5.5.5 and [40]. For this reason it was not clear, whether it is possible to change the thermometer temperature via this weak coupling on the time scale of a typical pulse ($\approx 10\,\text{ms}$) by warming up the heater. It turned out that the operating point of the thermometer **can be controlled** by heater pulses, too.[3]

To illustrate this, in figure 6.4 a normalized heater pulse of light detector X, whose thermometer and heater are separated, can be seen. In addition, pulses caused by direct hits in the light absorber and scintillation light absorptions are shown. Fitting these pulses by the solution derived in the model of chapter 5 results in the parameters listed in table 6.1. In case of the normalized heater pulses the fit quality was not as good as in the two other cases. This is taken into account by giving a parameter range. Physically, this is a hint that the model's solution, and therefore the model's assumptions, is not as good for heater pulses as for particle pulses.

In all three cases, τ of table 6.1 is connected to the rise time of the thermometer pulse. In case of direct absorber hits, it is equal to the non-thermal phonon life time τ_N. In the case of a scintillation event, it reflects the scintillation time constant τ_S of $CaWO_4$ at ultra-low temperatures. Combined with the non-thermal phonon life time τ_N, it determines the rise time of the pulse. In the case of heater pulses, τ is the rise time which is given by the coupling of the heater to the thermometer. **The time constant with which energy is transported from the heater into the thermometer, is in this case of the order of a few milliseconds.**

To get a better understanding of the physics in the light detector during a heater pulse, the **temperature changes** of **thermometer** and **heater** can be derived and compared:

---

[3] In RUN 32 of the CRESST-II experiment at Gran Sasso two light detectors, X and Yoichiro, have been operated with separated heaters.

6.1 Experiments

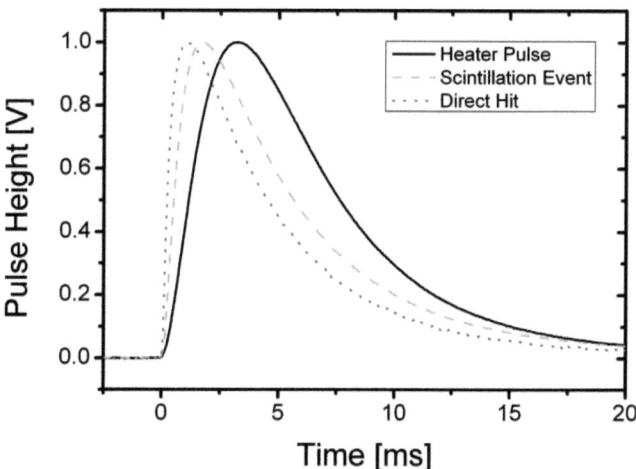

Figure 6.4: In this plot three averaged and normalized detected pulses of light detector X, whose thermometer and heater are separated, are shown. For direct hits the energy deposition takes place on a negligibly short time scale compared to the rise time of the pulse. In this case the rise time is given by the life time of the non-thermal phonons, cp. chapter 5. The rise time of a scintillation event is influenced, beside the non-thermal phonon life time, by the scintillation decay time of $CaWO_4$ at ultra-low temperatures. In the case of a heater pulse, the rise time of the thermometer pulse is given by the coupling of the heater to the thermometer.

X	A [%]	τ_+ [ms]	τ_- [ms]	τ [ms]
Direct Hit	98	4.13	27.11	0.40
Scintillation Event	98	4.52	28.37	0.48
Heater Pulse	99	2.5 - 4.5	15 - 30	2 - 2.5

Table 6.1: In this table, the fitted parameters of the three different averaged pulses of light detector X, which are shown in figure 6.4, are given. For the time constants of the heater pulse fit only ranges can be given due to the reduced fit quality. In all three cases, the fast component dominates the total measured pulse. This is reflected by A, which is close to one. The fast time constant (τ_+) represents the decay time of the pulses. This time constant is very similar in all cases. The slow time constant (τ_-) describes the negligibly slow component decay time. The third time constant (τ) is in the case of direct hits the rise time of the pulse, which is equal to the non-thermal phonon life time. In case of the scintillation events, τ represents the decay time of the $CaWO_4$ scintillation light output. Together with the non-thermal phonon life time (time constant τ of the direct hits), it determines the rise time of the scintillation event pulses. For heater pulses, the time constant τ represents the scale on which the energy is transported from the heater to the thermometer.

- To derive the **temperature change of the thermometer** during a heater pulse, the heater calibration can be used: For light detector X the thermometer warms up by the same amount in case of a 10 V heater pulse and a 2.11 MeV energy deposition in the target crystal caused by gamma radiation:
$$E_{dep} = 2.11\,\text{MeV}$$
The fraction of this energy which is absorbed by the light absorber of X is $p\,q = 1.23\,\%$ [58]:
$$E_{LD} = p\,q \cdot E_{dep} = 26\,\text{keV}$$
As seen in table 5.1, pulses of light detector X consist to 98 % of the fast component A_+. I.e. the fraction ε^* of non-thermal phonons, which is estimated to be about 5 %, cp. section 5.5.5, heats up the thermometer's heat capacity:
$$\boxed{\Delta T_T \approx \frac{\varepsilon^* \, E_{LD}}{C_T} \approx 1.6\,\text{mK}}$$
For an energy deposition of 2.11 MeV in the target, **the thermometer warms up by a few millikelvin.**

- This temperature rise of the thermometer is equal for heater pulses of 10 V. To **derive the temperature change of the heater** during such a heater pulse, it is assumed that the total amount of energy is deposited first in the heater, afterwards this energy distributes over the system.[4]
The energy deposited by a 10 V heater pulse in the gold heater is given as:[5]
$$\Delta E_{Htr}^{Pulse} = \int_0^\infty P_{Htr}^{Pulse} \approx \frac{1}{4}\,\text{fJ}$$
The heat capacity of the gold heater can be derived with appendix A:[6]
$$C(T_{Htr}^{Pulse}) = \nu \cdot \gamma_{Au} \cdot T_{Htr} = \nu \cdot \gamma_{Au} \cdot \left(\frac{T_{Htr}^{Pulse} + T_{Htr}^0}{2}\right)$$
T_{Htr}^0 is the unknown temperature of the heater before the pulse. T_{Htr}^{Pulse} is the maximum temperature of the heater during a pulse.
The temperature rise of the heater during a pulse is then given as:
$$T_{Htr}^{Pulse} - T_{Htr}^0 = \frac{\Delta E_{Htr}^{Pulse}}{C(T_{Htr}^{Pulse})} \approx \sqrt{0.137\,\text{K}^2 + {T_{Htr}^0}^2} - T_{Htr}^0$$

---

[4] The current of the heater pulse decays exponentially with a time constant of 0.973 ms, whereas the response time of the thermometer is between 2 - 2.5 ms. For this reason, above assumption can only be a rough approximation.

[5] The maximum heater current is $I_0^2 = 1.24\,\mu\text{A}^2$; the heater resistance is estimated to be $R_{Htr} \approx 0.2\,\Omega$.

[6] $\nu \approx 5 \cdot 10^{-12}$ mol and $\gamma_{Au} = 0.729\,\frac{\text{mJ}}{\text{mol}\,\text{K}^2}$

6.1 Experiments

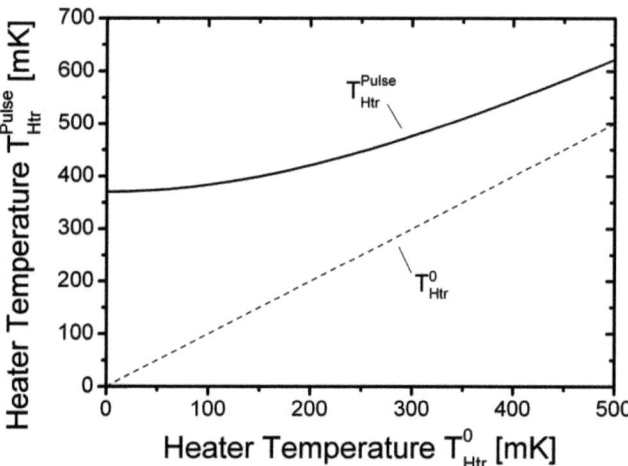

Figure 6.5: Depending on the unknown heater temperature before a pulse (T_{Htr}^0) the maximum temperature (T_{Htr}^{Pulse}) of this heater during a heater pulse of 10 V is shown. Depending on the heater temperature the temperature rise during a pulse is in the range of 100 - 350 mK.

Depending on the initial heater temperature T_{Htr}^0, plot 6.5 shows the maximum temperature of the heater. The temperature rise is given as the difference in between. It can be seen that the maximum heater temperature is expected to be larger than 350 mK and the temperature rise of the heater is in the range:

$$\boxed{\Delta T_{Htr} \approx 100 - 350\,\text{mK}}$$

The temperature rise of the heater for a pulse is about **two orders of magnitude** larger than the thermometer response. Due to the strong temperature dependence of the electron-phonon coupling (T^5) the energy deposited in the heater can be transferred relatively quickly through the phonon systems to the thermometer's electrons. This can explain the determined response time of the light detector X of about 2 ms.

> Light detector structures
> with separated thermometer and heater
> can be operated.

6.2 Discussion of Modifications

6.2.1 Thermometer Size

This section will discuss the influence of the thermometer size on the mean measured signal. Additionally to this, the measurement uncertainty in form of noise has to be taken into account for the influence on the finally relevant widths of the bands.

As seen in equation (4.1), the thermometer temperature rise depends on the energy transported into the thermometer and the thermometer's heat capacity:

$$\Delta T_T^{dep}(T) = \frac{r\, E_{LD}}{C_T(T)} \tag{6.1}$$

This heat capacity can be expressed using the specific heat by:

$$C_T(T) =: c_T(T) \cdot m = c_T(T) \cdot \rho\, d\, A \tag{6.2}$$

m is the mass of the thermometer; ρ is the thermometer density; d is the thermometer thickness, and A is the thermometer area.
The **assumption** that only a small fraction of non-thermal phonons is absorbed by the thermometer structure[7]

$$\varepsilon^* \ll 1 \tag{6.3}$$

results in the energy transported into the thermometer to be proportional to its area:

$$r = r(A) \approx r' \cdot A \tag{6.4}$$

Expressing the thermometer temperature rise by the specific heat and using this assumption leads to:

$$\Delta T_T^{dep}(T) = \frac{r'\, E_{LD}}{c_T(T) \cdot \rho\, d} \tag{6.5}$$

> **As long as the fraction of non-thermal phonons absorbed by the thermometer is small ($\varepsilon^* \ll 1$), the temperature rise of the thermometer is independent of the thermometer area A.**

As seen in section 5.5.5, in case of non-thermal phonon life times of $\tau_N \gtrsim 0.3\,\mathrm{ms}$, the thermometer temperature rise is dominated by the fast component A_+. This component is the overheating of the thermometer compared to the dynamic state

---
[7]Determinations of ε^* show that it is below 0.1, cp. section 5.5.5.

6.2 Discussion of Modifications

temperature after adjusting and it s caused by the non-thermal phonon absorption. It is expected that the absorption of non-thermal phonons is so strong that it is nearly independent of the thermometer thickness [40]:

$$\frac{\partial r'}{\partial d} = 0 \qquad (6.6)$$

As long as this is the case, the following holds for the temperature rise of the thermometer during a pulse:

$$\Delta T_T^{dep}(T) = \frac{r' E_{LD}}{c_T(T) \cdot \rho\, d} \sim \frac{1}{d} \qquad (6.7)$$

As long as the fraction of non-thermal phonons absorbed by the thermometer is nearly independent of the thickness d, the temperature rise of the thermometer is inversely proportional to the thermometer thickness d.

A maximization of the thermometer temperature rise during a pulse could be reached by a thinner thermometer film. It should be noted that tungsten of 30 nm in case of so-called superconducting strips, see appendix C.1, and of 40 nm in [63] got superconducting and can therefore be used as thermometer. This thickness is about one order of magnitude thinner than the thermometers with a thickness of 200 nm used up to now.

However, it should be taken into account that finally the signal output measured is:

$$\Delta U_{out} \sim \Delta T_T^{dep} \cdot I_T \qquad (6.8)$$

This signal is not only proportional to the thermometer's temperature rise, it is proportionally to the read out current of the thermometer I_T, too. For this reason the up to now unknown dependence of the possible read out current on the thermometer thickness has to be taken into account. Additionally, the widths of the bands in the light-phonon-plane, which are of interest, depend on the noise which is influenced by the read out current and the thermometer resistance, too.

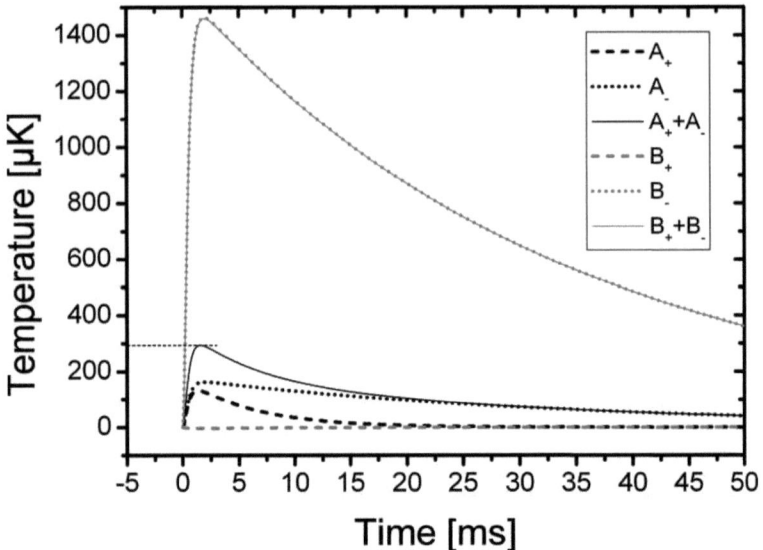

Figure 6.6: The fitted thermometer ($A_+ + A_-$) and derived absorber ($B_+ + B_-$) temperature rises of light detector Steven after a 122 keV gamma absorption in the target crystal. For simplification the components are only marked by A_+ and so forth. The time dependent brackets, which are noted in the equations (B.18) and (B.19), are left out. The detected maximum thermometer temperature rise of 294 µK is marked by a horizontal line. The fast (A_+) and the slow (A_-) component add up to the thermometer temperature ($A_+ + A_-$). This holds in the same way for the absorber components, B_+ and B_-, which sum up to the absorber's temperature ($B_+ + B_-$). It can be seen that the absolute value of the fast absorber component (B_+) is very small and negative (undercooling). This small amount results in the fact that the slow component dominates the absorber's temperature: $B_- \approx B_+ + B_-$

6.2.2 Maximization of ΔT_e

Different strategies can be followed to maximize the thermometer temperature rise ΔT_T, which has been derived in the model of chapter 5. Two of these will be discussed in this section.

The first possibility aims to **enlarge the overheating**, i.e. the goal is to increase the fast component $\boldsymbol{A_+}$. In this way the thermometer's dynamic state after adjusting (A_-) within the fast time constant τ_+ can be neglected.

For this case, it is important that the number of absorbed non-thermal phonons per thermometer volume is maximized. Thus, the non-thermal phonon decay time in the absorber (τ_A), which is dominated by inelastic surface scatterings, should be as long as possible. Due to the strong electron-phonon coupling of non-thermal phonons in the thermometer, for the power flow the thermometer area counts

6.2 Discussion of Modifications

instead of the volume, cp. section 4.2.3. The thickness of the absorber wafer and therefore its heat capacity can be chosen larger, on the contrary. This guarantees a homogeneous non-thermal phonon density which is not reduced close to the thermometer due to the absorptions. In order not to reduce the overheating by the time dependence, the fast time constant τ_+ should be much larger than the rise time of the thermometer pulse which is determined by the non-thermal phonon life time τ_N and the scintillation time of the target at ultra-low temperatures τ_S. As seen in section 5.3.4, the possible increase of the signal due to phonon collectors does not seem to be reached. For this kind of maximization, optimized phonon collectors could contribute significantly.

The second strategy maximizes the **slow thermometer parameter A_-**. This parameter is equal to the thermometer temperature after the fast time constant τ_+. This is the time constant on which the temperatures of thermometer and absorber adjust to each other, cp. section 5.5.4.

$$G_{AT} \gg G_{BT}^*, G_{AB}$$

Additionally, the heat capacities of the absorber and thermometer should be as small as possible, see discussion above. Silicon absorption wafers of about 50 μm thickness (40 mm in diameter) turn out to be most feasible due to their elastic property.

The potential of this strategy can be seen in figure 6.6. In this picture the fitted time dependent thermometer pulse of light detector Steven can be seen $(A_+ + A_-)$. It consists of the fast (A_+) and slow (A_-) temperature component, which have been derived in section 5.5.5. In the same way the fast (B_+) and slow (B_-) components of the absorber temperature $(B_+ + B_-)$ have been derived.

It can be seen that the absorber warms up significantly more than the thermometer. This is not surprising, since about 95 % of the non-thermal phonons decay in the absorber and warm it up. As seen in section 5.5.5, the absorber heat capacity can be similar to the thermometer's heat capacity. An adjustment of these two temperatures could enlarge the thermometer temperature significantly.

Chapter 7

Conclusions and Perspectives

It has been shown in chapter 2 that the region in the light-phonon-plane, where events show up, depends on the incident particle. These different areas have the shape of bands. Each mean band position is fixed by the respective light-to-phonon ratio detected. The widths of the bands are determined mainly by the energy resolution of the light channel. The knowledge of the bands in the light-phonon-plane is essential for WIMP detection.

The positions of these bands relative to each other can be described by the so-called quenching factors, see chapter 3. It has been shown in this work that the quenching factors are given by the energy dependent energy loss per path length, $dE/dx\,(E)$, and that the light production per path length is correlated to the energy loss, $dL/dx\,(dE/dx)$. The energy loss is naturally given and connected to the target material. Thus the only way to change the relative band positions can be a change of dL/dx. If this is connected to the target material, too, then the relative band positions will be fixed and can only be changed by a change of the target material.

The width of the band is the second factor which determines the areas covered by the different bands and, in this way, the WIMP sensitive energy range. The parameters and physical properties which mainly influence the widths of the bands are determined in chapter 4 and chapter 5. These are the quantities which determine the energy resolution of the light channel. Independently of the choice of the target material, an improvement of the reflecting housing or of the resolution of the light detector, which is the main part of the light channel, will reduce the band widths in any case. Thus, independently of the target material, the WIMP sensitivity can be increased. In case of $CaWO_4$ targets, the overlapping areas of the tungsten band with other bands are important and can be efficiently reduced by an improvement of the energy resolution of the light channel for energies $\lesssim 10\,\text{keV}$.

An improvement of the energy resolution of the light detector is thus essential. The parameters which can be changed were determined in the chapters 4 and 5 of this work. In chapter 6, different changes have been discussed and realized.

Furthermore, more changes can be applied to the light channel, which was, up to now, mainly unoptimized:

Light and Phonon Transport: One important property of the light channel is that the first part of the energy transport is in form of photons, whereas the second part is in form of phonons. Therefore, the ratio of the energy transport in form of photons to the one in form of phonons can be optimized. This can be done by the minimization of the total energy loss on the transport path from the target crystal to the light detector thermometer.
The photon transport is determined, beside the target crystal, by the reflecting housing, the light absorber size and the absorption probability. On the other hand, the phonon transport is determined by the non-thermal phonon life time, the coupling of the absorber to the thermometer, and the thermometer geometry. A change of one of these parameters involves a different optimal ratio of these two energy transports (photon and phonon). For example, the annealing of the light absorber might change the surface properties and, in this way, the life time of the non-thermal phonons, which is determined by the inelastic scatterings at the absorber surface. A longer non-thermal phonon life time can reduce the energy loss during the phonon transport. Therefore, the transport via non-thermal phonons may be improved and the loss in the transport via photons can be reduced by reducing this transport length. Thus, the total energy loss due to the transfer is reduced.

Phonon Collectors: As seen in section 5.3.4, the phonon collectors do not seem to work efficiently. The reason for this could be that the thermometer is much thinner than the phonon collectors (5:1), which may reduce the transmission probability from the collectors into the thermometer. Thinner collector films could solve this problem. As comparison, in [40], where quasi particle diffusion could be observed, the collector film was only twice as thick as the thermometer.
Furthermore, the shape and size of the phonon collectors can be optimized depending on the diffusion properties and life time of the quasi particles.

Transition Temperature: Another aspect, which influences the energy resolution of the light detector is the transition temperature of the thermometer. Since the heat capacities of thermometer and absorber are temperature dependent and inversely proportional to the temperature rise after an energy deposition, a lower transition temperature can improve the energy resolution. A reduction of the transition temperature can be reached via the proximity effect. Thus, for example, a thin gold layer on top of the thermometer could reduce the transition temperature. Another possibility might be a surface treatment of the absorber below the thermometer. In average, a surface roughened by an argon ion gun seems to increase the transition temperature, due to the preferred polycrystalline thin film growing. However, the operation temperature has to remain in the range which can be reached with the cryostat.

Transition Slope: The steeper the transition of a thermometer from the superconducting to the normal-conducting state is, the better is the expected energy resolution of the light channel. Therefore, steep transitions are preferred. A way to increase the steepness could be to reduce the influence of the change of the critical temperature perpendicular to the read out bias current. This can be realized by perpendicular superconducting (e.g. Al) strips on the thermometer.

Another possibility for improving the transition slope could be the annealing of the thermometer film. In this way a more homogeneous film can be produced and a more homogeneous transition temperature can be expected. Furthermore, much smaller thermometers could be realized. The number of absorbed non-thermal phonons per area would not decrease, whereas, due to the small size, the relative transition temperature homogeneity could be increased.

Readout Resistance Ratio: As seen in chapter 4.2, the mean detected signal is inversely proportional to the sum of the thermometer and shunt resistances. On the other hand, to optimize these two resistance values, the noise introduced by them has to be taken into account. It should be noted that the thermometer resistance can be changed without changing its geometry and heat capacity. This is possible since a roughening of the surface below the thermometer results in polycrystalline thermometer films. The resistances of these films are typically about three times larger, whereas the geometry and the heat capacity stays the same.

In summary, a large space of improvements for the energy resolution of the light channel still exists. An improvement of this channel results in narrower bands in the light-phonon-plane and therefore in an improved WIMP identification sensitivity of CRESST-II. For this reason, further research and development of the light channel is essential.

Appendix A

Heat Capacities

The heat capacity C is the property which characterizes its temperature change ∂T following a heat change δQ:

$$C = \frac{\delta Q}{\partial T}$$

If, for example, energy is absorbed, it will distribute equally over all internal degrees of freedom of the system and warm it up in this way.

In general the heat capacity depends on the temperature T:
At temperatures much smaller than the Debye temperature, $T \ll \Theta_D$, the heat capacity of the phonon system of a solid body, which is equal to the total heat capacity in case of a **non-metal**, can be described very well by the Debye model [64]:

$$C^p(T) = A \cdot \nu \cdot \frac{12\pi^4}{5} \cdot N_A \cdot k_B \cdot \left(\frac{T}{\Theta_D}\right)^3 =: A \cdot \nu \cdot B \cdot T^3$$

A is the number of atoms per unit cell, ν is the number of mol of the system, $N_A = 6.022 \cdot 10^{23}$ mol^{-1} is the Avogadro constant, $k_B = 1.38 \cdot 10^{-23}$ $\frac{\text{J}}{\text{K}} = 8.62 \cdot 10^{-5}$ $\frac{\text{eV}}{\text{K}}$ is the Boltzmann constant, Θ_D is the Debye temperature and B is a constant.

The heat capacity of a **metal** has, in addition to the previous phonon system, a contribution of the electron system. This contribution can be described at temperatures $T \ll T_F$ very well by the Sommerfeld theory [65]:

$$C^e(T) = \nu \cdot \frac{\pi^2}{2} \cdot n_e \cdot k_B \cdot \frac{T}{T_F} =: \nu \cdot \gamma \cdot T$$

ν is the number of mol of the system, n_e is the number of conduction electrons per mol, $k_B = 1.38 \cdot 10^{-23}$ $\frac{\text{J}}{\text{K}} = 8.62 \cdot 10^{-5}$ $\frac{\text{eV}}{\text{K}}$ is the Boltzmann constant, T_F is the Fermi temperature of the electrons and γ is the Sommerfeld parameter.

At temperatures $T \ll 10$ K, the heat capacity contribution of the electrons in a metal is typically much larger than the one of the phonons. The latter can thus be neglected.

If the metal is a **superconductor** the heat capacity of the electrons changes below the critical temperature T_c. Above the critical temperature the heat capacity is

$$C^e(T > T_c) = \nu \cdot \gamma \cdot T$$

as for a non-superconductor. The transition from the normal-conducting phase into the superconducting phase is a second order phase transition with the consequence of a jump in the heat capacity at the critical temperature T_c. For metals with weak electron phonon interaction, as it is the case for tungsten [66], BCS theory predicts the size of the jump [65][67]:

$$\Delta C^e(T_c) = 1.43 \cdot \nu \cdot \gamma \cdot T_c = 1.43 \cdot C^e(T_c)$$

Well below the critical temperature, BCS theory predicts that the heat capacity of the electron system of a superconductor decreases exponentially [64][68]:

$$C^e(T \ll T_c) \sim exp\left(-D \cdot \frac{T_c}{T}\right)$$

D is a constant.

The heat capacities of the different **light detector components** will be calculated in the following with the help of the previous information and literature values:

Phonon Heat Capacity of the Light Absorber C_A^p

The absorber is mainly a sapphire wafer with a diameter of 40 mm and a thickness of 0.46 mm. The Debye temperature of sapphire is $\Theta_D^{Al_2O_3} = 1041$ K [39]. From this the absorber temperature (T_A) dependent heat capacity can be calculated:

$$C_A^p(T_A) = 1.23 \frac{\text{eV}}{\text{mK}} \cdot \left(\frac{T_A}{\text{mK}}\right)^3$$

Phonon Heat Capacity of the Thermometer C_T^p

The light detector thermometer is a 0.45 mm × 0.3 mm × 200 nm thin tungsten film with a Debye temperature of $\Theta_D^W = 383$ K [69]. The phonon heat capacity is:

$$C_T^p(T_T) = 0.6 \cdot 10^{-6} \frac{\text{eV}}{\text{mK}} \left(\frac{T_T}{\text{mK}}\right)^3$$

Electron Heat Capacity of the Thermometer C_T^e

The Sommerfeld parameter of tungsten is $\gamma_W = 1.008$ mJ/mol·K² [69]. The electron heat capacity of the tungsten thermometer for the normal-conducting and completely super-conducting[1] case, respectively, is given as:

$$C_T^e(T_T) = X \cdot 17.86 \frac{\text{eV}}{\text{mK}} \cdot \frac{T_T}{\text{mK}} \quad \begin{cases} X = 1 & \text{normal-conducting} \\ X = 2.43 & \text{super-conducting} \end{cases}$$

---
[1] Completely super-conducting implies here that the thermometer temperature is still equal to the critical temperature.

In figure 5.8 the calculated electron heat capacity of the thermometer C_T^e and the phonon heat capacity of the absorber C_A^p are shown as a function of their temperatures.[2] The phonon heat capacity of the thermometer C_T^p is at least five orders of magnitude smaller.

---

[2] For simplification C_T^e and C_A^p are only signed as C_T and C_A in figure 5.8, see chapter 5.

Appendix B

Solution of the Two Coupled Differential Equations

In this chapter the approach of the solution for the two coupled differential equations of the model from chapter 5 will be shown.

The equations to be solved are given in section 5.3.6 in (5.21) and (5.22):

$$C_T \cdot \Delta \dot{T}_T + (G_{AT} + G_{BT}^*) \cdot \Delta T_T - G_{AT} \cdot \Delta T_A = P_T \tag{B.1}$$

$$C_A \cdot \Delta \dot{T}_A + (G_{AT} + G_{AB}) \cdot \Delta T_A - G_{AT} \cdot \Delta T_T = P_A \tag{B.2}$$

The power inputs P_T and P_A are given in equation (5.15) and (5.16), respectively:[1]

$$P_T(t) = \varepsilon^* \cdot E_{LD} \cdot \frac{1}{\tau_N} e^{\left(-\frac{t}{\tau_N}\right)} \cdot \Theta(t)$$

$$P_A(t) = (1 - \varepsilon^*) \cdot E_{LD} \cdot \frac{1}{\tau_N} e^{\left(-\frac{t}{\tau_N}\right)} \cdot \Theta(t)$$

In the following the Θ function, which defines the point in time of the energy deposition in the light detector, will be left out for simplicity. I.e. for $t < 0$ the system is in steady state, for $t \geq 0$ the following ansatz describes the system's relative temperatures:

$$\Delta T_T = A_+ \left(e^{-\frac{t}{\tau_+}} - e^{-\frac{t}{\tau_N}} \right) + A_- \left(e^{-\frac{t}{\tau_-}} - e^{-\frac{t}{\tau_N}} \right) \tag{B.3}$$

$$\Delta T_A = B_+ \left(e^{-\frac{t}{\tau_+}} - e^{-\frac{t}{\tau_N}} \right) + B_- \left(e^{-\frac{t}{\tau_-}} - e^{-\frac{t}{\tau_N}} \right) \tag{B.4}$$

---

[1] This is the case for $\tau_S \ll \tau_N$. In case of $\tau_S \gg \tau_N$ the time constant τ_N has to be exchanged by τ_S, cp. section 5.3.5. Otherwise ($\tau_S \approx \tau_N$) both time constants have to be taken into account, see equation (5.17) and (5.19), which changes the ansatz as can be seen in equation (B.18) and (B.19). For all these cases the following solution parameters are the same. In the following the influence of τ_S will be neglected for clarity.

Inserting this ansatz into equation (B.1) gives three constraints; one for each of the linear independently exponential functions:

$$e^{-\frac{t}{\tau_+}}: \qquad \left(G_{AT} + G_{BT}^* - \frac{C_T}{\tau_+}\right) A_+ = G_{AT} B_+ \qquad (B.5)$$

$$e^{-\frac{t}{\tau_-}}: \qquad \left(G_{AT} + G_{BT}^* - \frac{C_T}{\tau_-}\right) A_- = G_{AT} B_- \qquad (B.6)$$

$$e^{-\frac{t}{\tau_N}}: \quad \left(\frac{C_T}{\tau_N} - G_{AT} - G_{BT}^*\right)(A_+ + A_-) + G_{AT}(B_+ + B_-) = \frac{\varepsilon^* E_{LD}}{\tau_N} \qquad (B.7)$$

Eliminating B_+ and B_- in (B.7) with (B.5) and (B.6) results in:

$$\left(1 - \frac{\tau_N}{\tau_+}\right) A_+ + \left(1 - \frac{\tau_N}{\tau_-}\right) A_- = \frac{\varepsilon^* E_{LD}}{C_T} \qquad (B.8)$$

Inserting the ansatz into the second equation (B.2) gives three more constrains:

$$e^{-\frac{t}{\tau_+}}: \qquad \left(G_{AT} + G_{AB} - \frac{C_A}{\tau_+}\right) B_+ = G_{AT} A_+ \qquad (B.9)$$

$$e^{-\frac{t}{\tau_-}}: \qquad \left(G_{AT} + G_{AB} - \frac{C_A}{\tau_-}\right) B_- = G_{AT} A_- \qquad (B.10)$$

$$e^{-\frac{t}{\tau_N}}: \quad \left(\frac{C_A}{\tau_N} - G_{AT} - G_{AB}\right)(B_+ + B_-) + G_{AT}(A_+ + A_-) = \frac{(1-\varepsilon^*) E_{LD}}{\tau_N} \qquad (B.11)$$

Eliminating A_+ and A_- in (B.11) with (B.9) and (B.10) results in:

$$\left(1 - \frac{\tau_N}{\tau_+}\right) B_+ + \left(1 - \frac{\tau_N}{\tau_-}\right) B_- = \frac{(1-\varepsilon^*) E_{LD}}{C_A} \qquad (B.12)$$

Including (B.5), (B.6), and (B.8) into (B.12) gives A_+ as:

$$A_+ = \frac{1}{1 - \frac{\tau_N}{\tau_+}} \cdot \frac{1}{\frac{1}{\tau_-} - \frac{1}{\tau_+}} \cdot \left[\varepsilon^* \cdot \left(\frac{1}{\tau_-} - \frac{G_{AT} + G_{BT}^*}{C_T}\right) + (1-\varepsilon^*) \cdot \frac{G_{AT}}{C_A}\right] \cdot \frac{E_{LD}}{C_T} \qquad (B.13)$$

With (B.13) and (B.8) follows for A_-:

$$A_- = \frac{1}{1 - \frac{\tau_N}{\tau_-}} \cdot \frac{1}{\frac{1}{\tau_+} - \frac{1}{\tau_-}} \cdot \left[\varepsilon^* \cdot \left(\frac{1}{\tau_+} - \frac{G_{AT} + G_{BT}^*}{C_T}\right) + (1-\varepsilon^*) \cdot \frac{G_{AT}}{C_A}\right] \cdot \frac{E_{LD}}{C_T} \qquad (B.14)$$

In the same way B_+ and B_- can be derived: Substituting equations (B.9), (B.10), and (B.12) into (B.8) results in:

$$B_+ = \frac{1}{1-\frac{\tau_N}{\tau_+}} \cdot \frac{1}{\frac{1}{\tau_-}-\frac{1}{\tau_+}} \cdot \left[(1-\varepsilon^*) \cdot \left(\frac{1}{\tau_-} - \frac{G_{AT}+G_{AB}}{C_A}\right) + \varepsilon^* \cdot \frac{G_{AT}}{C_T}\right] \cdot \frac{E_{LD}}{C_A} \tag{B.15}$$

With (B.15) and (B.12) follows for B_-:

$$B_- = \frac{1}{1-\frac{\tau_N}{\tau_-}} \cdot \frac{1}{\frac{1}{\tau_+}-\frac{1}{\tau_-}} \cdot \left[(1-\varepsilon^*) \cdot \left(\frac{1}{\tau_+} - \frac{G_{AT}+G_{AB}}{C_A}\right) + \varepsilon^* \cdot \frac{G_{AT}}{C_T}\right] \cdot \frac{E_{LD}}{C_A} \tag{B.16}$$

The two time constants have to fulfill the same condition. It follows from (B.5) and (B.9), (B.6) and (B.10), respectively:

$$\tau_{+,-} = \frac{2}{a \pm \sqrt{a^2 - 4b}} \tag{B.17}$$

a and b are defined as:

$$a := \frac{G_{AT}+G_{BT}^*}{C_T} + \frac{G_{AT}+G_{AB}}{C_A} = \frac{1}{\tau_+} + \frac{1}{\tau_-}$$

$$b := \frac{G_{AT}G_{BT}^* + G_{AT}G_{AB} + G_{BT}^*G_{AB}}{C_T C_A} = \frac{1}{\tau_+} \cdot \frac{1}{\tau_-}$$

$$\sqrt{a^2 - 4b} = \sqrt{\left(\frac{G_{AT}+G_{BT}^*}{C_T} - \frac{G_{AT}+G_{AB}}{C_A}\right)^2 + 4\frac{G_{AT}^2}{C_A C_T}} = \frac{1}{\tau_+} - \frac{1}{\tau_-}$$

In the ansatz (B.3) and (B.4) six parameters are free: $A_+, A_-, B_+, B_-, \tau_+$ and τ_-. With the solution of the two equations (B.1) and (B.2) these six parameters are connected in an unique way with the physical properties of the light detector: $G_{BT}^*, G_{AB}, G_{AT}, C_A, C_T, \tau_N, \varepsilon^*$ and E_{LD}.

In case of $\tau_S \approx \tau_N$, the ansatz contains both time constants and changes to:

$$\Delta T_T = A_+ \left(\frac{\tau_+}{\tau_+ - \tau_S}\left(e^{-\frac{t}{\tau_+}} - e^{-\frac{t}{\tau_S}}\right) - \frac{\tau_N}{\tau_S - \tau_N}\left(e^{-\frac{t}{\tau_S}} - e^{-\frac{t}{\tau_N}}\right)\right)$$

$$+ A_- \left(\frac{\tau_-}{\tau_- - \tau_S}\left(e^{-\frac{t}{\tau_-}} - e^{-\frac{t}{\tau_S}}\right) - \frac{\tau_N}{\tau_S - \tau_N}\left(e^{-\frac{t}{\tau_S}} - e^{-\frac{t}{\tau_N}}\right)\right) \tag{B.18}$$

$$\Delta T_A = B_+ \left(\frac{\tau_+}{\tau_+ - \tau_S}\left(e^{-\frac{t}{\tau_+}} - e^{-\frac{t}{\tau_S}}\right) - \frac{\tau_N}{\tau_S - \tau_N}\left(e^{-\frac{t}{\tau_S}} - e^{-\frac{t}{\tau_N}}\right)\right)$$

$$+ B_- \left(\frac{\tau_-}{\tau_- - \tau_S}\left(e^{-\frac{t}{\tau_-}} - e^{-\frac{t}{\tau_S}}\right) - \frac{\tau_N}{\tau_S - \tau_N}\left(e^{-\frac{t}{\tau_S}} - e^{-\frac{t}{\tau_N}}\right)\right) \tag{B.19}$$

For the solution parameters nothing changes, except there is one more independent parameter τ_S. This time constant is the scintillation decay time of $CaWO_4$ at ultra-low temperatures.

Appendix C
Practical Remarks

C.1 Superconducting Cooling Strip

The connection of the light detector thermometer to the heat bath is mainly given by a long and thin ($\approx 30\,\text{nm}$) gold strip. Such a strip can be seen in figure 4.3(a). Due to the noble properties of gold their adhesion on the absorber surface (Al_2O_3) is relatively weak. To improve this, a thin ($\approx 30\,\text{nm}$) tungsten layer can be used in between. The adhesion of tungsten on sapphire (Al_2O_3) is significantly stronger than the one of gold, and the adhesion of gold on tungsten is significantly stronger than on Al_2O_3 as well.
Using a thin tungsten layer in between the sapphire (Al_2O_3) and the gold (Au) thus improves the adhesion of the cooling strip.

However, a drawback of this tungsten-gold bilayer is that it gets superconducting at ultra-low temperatures. Especially the normally used sputtered tungsten can get superconducting at a few Kelvin. The exact transition temperature of bi-layer depends on the lattice structure of the tungsten film and on the ratio of the gold to tungsten. In any case, both thin layers get superconducting at the same temperature due to the proximity effect.

If a metal gets superconducting the electrons will pair to Cooper-pairs. These Cooper-pairs do not contribute to the heat transfer anymore, i.e. these electrons of the tungsten-gold bilayer do not cool the thermometer anymore.

In figure C.1 the resistance of a gold strip is shown as a function of the temperature. It can be seen that at about $350\,\text{mK}$ the resistance drops to less than $50\,\%$. I.e. at this temperature the first electrons pair to Cooper-pairs. At about a tenth of the critical transition temperature ($T_c/10 \approx 35\,\text{mK}$) nearly all electrons are paired. Their cooling is suspended.[1] The thermometer cannot be cooled into the transition and cannot be operated.

For this reason no tungsten layer is used between the sapphire and the gold in CRESST-II any more. Only tungsten below the bond pad is left over as it is needed for bonding. To reduce the influence of this tungsten onto the cooling,

---
[1] This can be shown for example, for aluminum: At $T \leq T_c/10$ the thermal conductivity is reduced by about six orders of magnitude. The heat conductivity is provided by phonons [51].

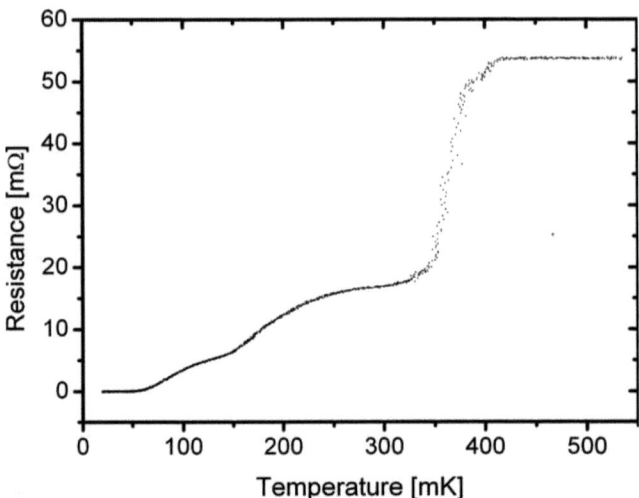

Figure C.1: In this figure the temperature dependent resistance of a gold cooling strip, including a sputtered tungsten layer, can be seen. At about 350 mK more than 50 % of the strip gets superconducting, i.e. at about 35 mK the electron cooling is completely suppressed. The cooling strip does not work anymore at lower temperatures and therefore the light detector cannot be operated.

evaporated tungsten left over from the thermometer production is used (instead of sputtered tungsten in a separate step). In this way the tungsten below the bond pad has about the same transition temperature as the thermometer and provides enough electrons for the energy transport.

After removing the tungsten below the gold at old light detectors and producing new light detectors without the tungsten interlayer, **29 out of 29** of these light detectors have been operated **successfully**.

C.2 Destroyed Gold Structure

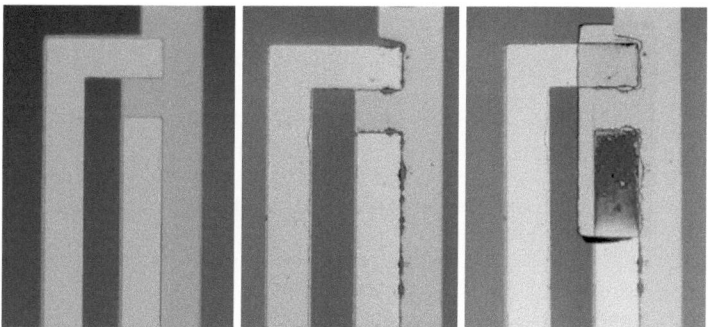

Figure C.2: On the left hand side, a typical part of the strip around the heater, which is the area in between the two aluminum ends, can be seen. The middle picture shows this part of another thermometer strip after destruction caused by electro static discharge. As can be seen, the gold at the border to the aluminum is removed. No electrical connection is given any more between the two aluminum lines. On the right hand side the same heater can be seen with an additional gold patch, which restores the full functionality.

C.2 Destroyed Gold Structure

In figure C.2 three pictures are shown, each showing the heater part of a light detector thermometer structure. Between the two aluminum lines there is a $40\,\mu m \times 40\,\mu m$ gold connection. This part is used as heater.[2] In the picture on the left hand side, a newly produced heater can be seen.

In the middle picture, gold at the border to the aluminum is destroyed. There is no electrical connection between the two aluminum heater lines anymore. The detector cannot be operated anymore due to the disconnected heater.

On the right hand side, the same heater can be seen again after patching. The area around the heater is covered by an additional gold structure. In this way the light detector can be operated again.

It is assumed, that electrostatic discharge (ESD) is the reason for the heater destruction (from the left to the middle picture). This could be confirmed in a dedicated experiment, where systematic charge carriers have been discharged over the structure with the same result [70].

---

[2] At ultra-low temperatures the aluminum is superconducting in contrary to the gold. Therefore a current from one aluminum line to the other dissipates heat only in the gold part.

C.3 Measurement of the Thermometer Temperature Increase During a Pulse

For detector research and development it is essential to know the properties of the detector, cp. chapter 4 and 5. Such properties are, for example, the collected energy in the thermometer $r \cdot E_{LD}$ after an energy deposition in the detector target, the heat capacity of the thermometer C_T, or the thermometer temperature increase ΔT_T^{dep} caused by a pulse.

These three properties are connected with each other as follows, see also equation (4.1):

$$\Delta T_T^{dep} \approx \frac{r E_{LD}}{C_T}$$

For getting information about one of these three properties, usually the knowledge of the **temperature increase of the thermometer** $\boldsymbol{\Delta T_T^{dep}}$ is necessary first. However, the direct measurement of this temperature change is not possible, due to the superposition of two concomitant effects: The bias current heating effect and the critical current effect.

Here it will be shown that in case of small electro-thermal feedback the temperature change of interest can be determined with a bath temperature sweep.

The consequence is that large bias currents, where the bias effects are also large, can be used for measurements. This provides a better signal since the signal output is proportional to the thermometer current, cp. equation (4.3):

$$\Delta U_{out} \sim I_T$$

Another consequence is that a small pulse approximation does <u>not</u> have to be used, since the method is independent of the transition characteristic.

In the following, two measurements will be considered and compared: First a measurement of a temperature pulse, where the thermometer is stabilized in the transition. Second a bath temperature sweep, where the thermometer temperature is changed via the bath temperature.

For measuring **temperature pulses** the thermometer of a detector is stabilized in the transition at a temperature $T = T_T^1$. This temperature is given by the sum of the bath temperature T_B and the bias current heating T_{Bias}^1 in this point:

$$T_T^1 = T_B + T_{Bias}^1 \tag{C.1}$$

After an energy collection, the thermometer is heated up by $\Delta T_T^{dep} \approx \frac{r E_{LD}}{C_T}$. This changes the bias current heating to T_{Bias}^2. The total temperature T_T^1 rises to T_T^2 accordingly:

$$T_T^2 = T_B + T_{Bias}^2 + \Delta T_T^{dep} \tag{C.2}$$

C.3 Thermometer Temperature Increase During a Pulse

The temperature change of the thermometer is then:

$$\Delta T_T := T_T^2 - T_T^1 = \Delta T_T^{dep} + \Delta T_{Bias}$$

$\Delta T_{Bias} := T_{Bias}^2 - T_{Bias}^1$, it is the change of the bias current heating. The property of interest is ΔT_T^{dep}, which is caused by the energy deposition. The thermometer temperature rises by ΔT_T instead.

This temperature change ΔT_T can be measured and for small pulses, where the transition slope m is constant, approximated as (cp. (4.3)):

$$\Delta U_{out} \sim m \cdot \Delta T_T$$

For large pulses the transition slope is not constant anymore, since critical current effects and the transition characteristic have an influence. Then above approximation is not valid and the voltage output of the SQUID can only be given as:

$$\Delta U_{out} = U_{out}^2(T_T^2) - U_{out}^1(T_T^1) \tag{C.3}$$

For small and large pulses, the bias heating effect and the critical current effect determine the unknown relation between the thermometer temperature change ΔT_T^{dep} of interest and the finally measured SQUID voltage change ΔU_{out}.

To obtain ΔT_T^{dep} from ΔU_{out} a **bath temperature sweep** can be used. Therefor the SQUID voltage output U_{out} is measured for all different bath temperatures T_B in the relevant range. Certainly, in this measurement the SQUID output measures the total thermometer temperature T_T, which is given here as:

$$T_T = T_B + T_{Bias}$$

However, in this sweep, the bath temperature T_B is now changed continuously, i.e. all three temperatures change, since the bias current heating depends on the thermometer temperature.

Nevertheless, the bath temperature sweep contains also the values U_{out}^1 and U_{out}^2 as before in equation (C.3), which correspond to the same T_T^1 and T_T^2 as in equation (C.1) and (C.2), respectively:

$$\Delta U_{out} = U_{out}^2(T_T^2) - U_{out}^1(T_T^1)$$

Since the temperature change is caused by the bath sweep the respective temperatures are given here as:[3]

$$T_T^1 = T_B^1 + T_{Bias}^1 \tag{C.4}$$
$$T_T^2 = T_B^2 + T_{Bias}^2 \tag{C.5}$$

---
[3] In fact, T_{Bias}^2 is only similar to the one of the pulse measurement. The reason for this is, that in the case of the pulse, the bias heating change changes the heating of the thermometer only for a short time, τ_N, so that this temperature cannot arrange to the equilibrium temperature. However, in case of small electro-thermal feedback, as it is usually the case in CRESST, this effect is expected to be small.

The temperature shifts T^1_{Bias} and T^2_{Bias} caused by the bias current are the same as in the pulse measurement, since these shifts depend only on the thermometer temperature (cp. equation (5.9)):

$$T_{Bias}(T_T)$$

The comparison of the sweep (equation (C.1) and (C.2)) and the pulse measurement from above (equation (C.4) and (C.5)) then shows:

$$T_B + T^1_{Bias} = T^1_B + T^1_{Bias}$$
$$\Delta T^{dep}_T + T_B + T^2_{Bias} = T^2_B + T^2_{Bias}$$

And it follows:

$$\boxed{\Delta T^{dep}_T = \Delta T_B}$$

Where $\Delta T_B := T^2_B - T^1_B$ is the change of the bath temperature.

The temperature change caused by the energy deposition ΔT^{dep}_T is equal the measured bath temperature change ΔT_B, which yields the same ΔU_{out}.

With the help of a bath temperature sweep the temperature rise of the thermometer can thus be determined. Or in other words: The thermometer will warm up in the same way, regardless of whether it is caused by an energy deposition in the absorber or by an increase of the bath temperature. For each energy deposition there is an equivalent change in the bath temperature, which warms up the thermometer film by the same amount. Afterwards, in both cases the bias influence on the measured voltage will be the same.

Up to now only relative temperature changes have been considered. On the other side, the **absolute temperature** of the thermometer T_T can be of interest, for example for the determination of the thermometer heat capacity C_T. In this context should be noted that the bias heating effect has not only a dynamic influence on the thermometer temperature, it also has a constant one. In the operating point R_T, the bias current heating keeps a constant offset T_{Bias} between the bath temperature and the thermometer temperature:

$$T_T = T_B + T_{Bias}$$

Because of that the absolute temperature T_T is not known.
The shift caused by the constant bias current heating can be estimated with a series of measurements, where the influence of this effect decreases continuously. Therefor the sweep has to be measured with smaller and smaller currents, so that it can be extrapolated to a zero current measurement.

However, finally the limitation will be the uncertainty of the bath thermometer calibration.

Appendix D

Notation

In this thesis the following notation is used:

$$A_x^y(z)$$

- A: temperature T, resistance R, heat capacity C, thermal coupling G, energy E, current I, phonon/quasi-particle decay time τ, voltage U, atomic mass number A

- x: absorber A, thermometer T, shunt/SQUID S, diffusion D, non-thermal N, output out, heat bath B, light detector LD, heater Htr, bias current $Bias$

- y: phonon p, electron e, modified $*$, steady state 0

- z: time dependence t, temperature dependence T

A	Atomic mass number
A_+, A_-, B_+, B_-	Parameters of the model solution
Al_3O_3	Sapphire
c	Calibration factor
$C_T^e(T)$	Temperature dependent electron heat capacity of the thermometer
$C_T(T)$	Temperature dependent heat capacity of the thermometer which is about $C_T^e(T)$
$C_T^p(T)$	Temperature dependent phonon heat capacity of the thermometer
$C_A^p(T)$	Temperature dependent phonon heat capacity of the light absorber
$C_A(T)$	Temperature dependent heat capacity of the light absorber which is about $C_A^p(T)$
$CaWO_4$	Calcium tungstate
d	Energy fraction transported via quasi particles into the thermometer
E	Measured energy
$\langle E \rangle$	Averaged measured energy
E_{dep}	In target deposited energy
E_{LD}	In light absorber absorbed energy
ε	Energy fraction absorbed by the thermometer
ε^*	Energy fraction absorbed by the thermometer with existing additional phonon collectors
ε_C	Energy fraction absorbed by the phonon collectors
Θ_0	Magnetic flux quantum
G_{AB}	Thermal coupling between absorber and heat bath
G_{AT}	Thermal coupling between absorber and thermometer
G_{AT}^{pp}	Phonon-phonon coupling part of G_{AT}
G_{BT}	Thermal coupling between heat bath and thermometer
G_{BT}^*	G_{BT} including the electro thermal feedback
G_{TT}^{ep}	Electron-phonon coupling part of G_{AT}
I_{Tot}	Total bias current
I_T	Thermometer bias current
I^S	SQUID/Shunt bias current
ΔI_S	Current change of the SQUID during a pulse
ΔI_T	Current change of the thermometer during a pulse
I_{Htr}	Heater bias current
m	Transition slope
m^*	Transition slope in the R-T-plane
n	Index of refraction
p	Part of deposited energy transformed into scintillation light
P_A	Power flow into the absorber
P_T	Power flow into the thermometer

P_{Bias}	Dissipated power of the thermometer current
P_{Bias}^0	Dissipated power of the thermometer current in the steady state
$P_{Bias}^{\Delta T}(t)$	Time dependent change of the dissipated power of the thermometer current during a pulse
P_{Htr}	Dissipated power of the heater current
Π	Parameter including material and geometry of a thermal coupling
q	Fraction of absorbed scintillation light
r	Fraction of absorbed energy transferred to the thermometer
R_{Htr}	Heater resistance
R_T	Thermometer resistance
ΔR_T	Thermometer resistance change during a pulse
R_S	Shunt resistance
T	Temperature
$T_T^e(t)$	Time dependent electron temperature of the thermometer
$T_T(t)$	Time dependent temperature of the thermometer
T_T^0	Thermometer temperature in the steady state
T_T^p	Phonon temperature of the thermometer
ΔT_T^{dep}	Thermometer temperature rise caused by the energy deposition
ΔT_T	ΔT_T^{dep} including the temperature change caused by the bias current change
$T_A^p(t)$	Time dependent phonon temperature of the absorber
$T_A(t)$	Time dependent temperature of the absorber
T_A^0	Absorber temperature in the steady state
T_c	Critical temperature
T_B	Bath temperature
ΔT_B	Bath temperature change during a temperature sweep
τ_A	Time scale on which non-thermal phonons decay caused by surface decays in the absorber
τ_D	Diffusion time of the quasi particles
τ_N	Decay time of the non-thermal phonons
τ_S	Scintillation time constant of $CaWO_4$
τ_T	Time scale on which non-thermal phonons decay caused by thermometer absorptions
τ_+, τ_-	Fast/Slow time parameter of the model solution
U_{out}	Output voltage of the SQUID
X	Describes the fraction of the thermometer which is superconducting
δx	Measurement uncertainty of x
$ZnWO_4$	Zinc tungstate

Bibliography

[1] E. Komatsu, J. Dunkley, MR Nolta, CL Bennett, B. Gold, G. Hinshaw, N. Jarosik, D. Larson, M. Limon, L. Page, et al. Five-Year Wilkinson Microwave Anisotropy Probe Observations: Cosmological Interpretation. *Astrophys. J. Suppl*, 180:330–376, 2009.

[2] E. Corbelli and P. Salucci. The extended rotation curve and the dark matter halo of M33. *Arxiv preprint astro-ph/9909252*, 1999.

[3] M. Mateo. The Kinematics of Dwarf Spheroidal Galaxies. *Arxiv preprint astro-ph/9701158*, 1997.

[4] RG Carlberg, HKC Yee, E. Ellingson, R. Abraham, P. Gravel, S. Morris, and CJ Pritchet. Galaxy cluster virial masses and Omega. *The Astrophysical Journal*, 462:32, 1996.

[5] NASA, A. Fruchter, ERO Team, STScI, and ST-ECF. A Cosmic Magnifying Glass. *Hubble Space Telescope Institute*, 1994.

[6] L.L.R. Williams, J.F. Navarro, and M. Bartelmann. The core structure of galaxy clusters from gravitational lensing. *The Astrophysical Journal*, 527(2):535–544, 1999.

[7] H. Hoekstra, Y. Mellier, L. Van Waerbeke, E. Semboloni, L. Fu, MJ Hudson, LC Parker, I. Tereno, and K. Benabed. First Cosmic Shear Results from the Canada-France-Hawaii Telescope Wide Synoptic Legacy Survey 1. *The Astrophysical Journal*, 647(1):116–127, 2006.

[8] A.D. Lewis, D.A. Buote, and J.T. Stocke. Chandra Observations of A2029: The Dark Matter Profile Down to below 0.01r in an Unusually Relaxed Cluster. *The Astrophysical Journal*, 586(1):135–142, 2003.

[9] G. Hinshaw, JL Weiland, RS Hill, N. Odegard, D. Larson, CL Bennett, J. Dunkley, B. Gold, MR Greason, N. Jarosik, et al. Five-Year Wilkinson Microwave Anisotropy Probe Observations: Data Processing, Sky Maps, and Basic Results. *The Astrophysical Journal*, 180(2):225–245, 2009.

[10] V. Springel, S.D.M. White, and C.S. Frenk. The Millennium Simulation Project. *Max Planck Society for the Advancement of Science Press and Public Relations Department*, 2005.

[11] V. Springel, S.D.M. White, A. Jenkins, C.S. Frenk, N. Yoshida, L. Gao, J. Navarro, R. Thacker, D. Croton, J. Helly, et al. Simulations of the formation, evolution and clustering of galaxies and quasars. *Nature*, 435(7042):629–636, 2005.

[12] G. Jungman, M. Kamionkowski, and K. Griest. Supersymmetric dark matter. *Physics Reports*, 267(5-6):195–373, 1996.

[13] C. Amsler, M. Doser, M. Antonelli, DM Asner, KS Babu, H. Baer, HR Band, RM Barnett, E. Bergren, J. Beringer, et al. Particle Data Group. *Phys. Lett. B*, 667(1):36, 2008.

[14] A. Birkedal, A. Noble, M. Perelstein, and A. Spray. Little Higgs dark matter. *Physical Review D*, 74(3):35002, 2006.

[15] H.C. Cheng, J.L. Feng, and K.T. Matchev. Kaluza-Klein dark matter. *Physical review letters*, 89(21):211301, 2002.

[16] P. Ivanov, P. Naselsky, and I. Novikov. Inflation and primordial black holes as dark matter. *Physical Review D*, 50(12):7173–7178, 1994.

[17] G. Raffelt. Axions. *Space Science Reviews*, 100(1):153–158, 2002.

[18] T. Hebbeker. Can the sneutrino be the lightest supersymmetric particle? *Physics Letters B*, 470(1-4):259–262, 1999.

[19] A. Bottino, F. Donato, N. Fornengo, and S. Scopel. Probing the supersymmetric parameter space by weakly interacting massive particle direct detection. *Physical Review D*, 63(12):125003, 2001.

[20] P.J.T. Leonard and S. Tremaine. The local Galactic escape speed. *Astrophysical Journal*, 353(Part 1), 1990.

[21] F. Donato, N. Fornengo, and S. Scopel. Effects of galactic dark halo rotation on WIMP direct detection. *Astroparticle Physics*, 9(3):247–260, 1998.

[22] R.H. Helm. Inelastic and elastic scattering of 187-MeV electrons from selected even-even nuclei. *Physical Review*, 104(5):1466–1475, 1956.

[23] JD Lewin and PF Smith. Review of mathematics, numerical factors, and corrections for dark matter experiments based on elastic nuclear recoil. *Astroparticle Physics*, 6(1):87–112, 1996.

[24] M. Cribier, B. Pichard, J. Rich, JP Soirat, M. Spiro, T. Stolarczyk, C. Tao, D. Vignaud, P. Anselmann, A. Lenzing, et al. The muon induced background in the GALLEX experiment. *Astroparticle Physics*, 6(2):129–141, 1997.

[25] A. Drukier and L. Stodolsky. Principles and applications of a neutral-current detector for neutrino physics and astronomy. *Physical Review D*, 30(11):2295–2309, 1984.

[26] P. Meunier, M. Bravin, M. Bruckmayer, S. Giordano, M. Loidl, O. Meier, F. Pröbst, W. Seidel, M. Sisti, L. Stodolsky, et al. Discrimination between nuclear recoils and electron recoils by simultaneous detection of phonons and scintillation light. *Applied Physics Letters*, 75:1335, 1999.

[27] J. Astrom, P. Di Stefano, F. Proebst, L. Stodolsky, and J. Timonen. Brittle fracture down to femto-Joules-and below. *Arxiv preprint arXiv:0708.4315*, 2007.

[28] E. Pantic. Performance of Cryogenic Light Detectors in the CRESST-II Dark Matter Experiment. *PhD Thesis, Technische-Universität-München*, 2008.

[29] Q. Kronseder. Measurement done at MPI. 2009/10.

[30] P. Huff. Messung der Lichtausbeute von Rueckstosskernen in CaWO4. *Diploma Thesis, Technische Universität München*, 2006.

[31] Q. Kronseder. In preparation. *Diploma Thesis, Ludwig-Maximilians-Universitat München*, 2010.

[32] J.F. Ziegler, JP Biersack, and M.D. Ziegler. *SRIM-The stopping and range of ions in matter*. SRIM Co. - http://www.srim.org, 2008.

[33] H.H. Anderson, M.J. Berger, H. Bichsel, J.A. Dennis, M. Inokuti, D. Powers, S.M. Seltzer, D. Thwaites, J. E. Turner, and D.E. Watt. ICRU Report 37 - Stopping Powers for Electrons and Positrons. 1984.

[34] R. Lang. Analysis done at MPI. 2008.

[35] M. Kiefer et al. Composite CaWO4 Detectors for the CRESST-II Experiment. *13th Workshop on Low Temperature Detectors*, 2009.

[36] J. Ninkovic. Investigation of CaWO4 Crystals for Simultaneous Phonon-Light Detection in the CRESST Dark Matter Search. *PhD Thesis, Technische Universität München*, 2005.

[37] I. Bavykina. Measurement done at MPI. 2008.

[38] F. Petricca. Dark Matter Search with Cryogenic Phonon-Light Detectors. *PhD Thesis, Ludwig-Maximilians-Universität München*, 2005.

[39] F. Pröbst, M. Frank, S. Cooper, P. Colling, D. Dummer, P. Ferger, G. Forster, A. Nucciotti, W. Seidel, and L. Stodolsky. Model for cryogenic particle detectors with superconducting phase transition thermometers. *Journal of Low Temperature Physics*, 100(1):69–104, 1995.

[40] J. Loidl. Diffusion und Einfang von Quasiteilchen. *PhD Thesis, Ludwig-Maximilian-Universität München*, 1999.

[41] G. Angloher, C. Bucci, P. Christ, C. Cozzini, F. Von Feilitzsch, D. Hauff, S. Henry, T. Jagemann, J. Jochum, H. Kraus, et al. Limits on WIMP dark matter using scintillating CaWO4 cryogenic detectors with active background suppression. *Astroparticle Physics*, 23(3):325–339, 2005.

[42] N.E. Phillips. Heat Capacity of Aluminum between 0.1 K and 4.0 K. *PHYSICAL REVIEW*, 114(3), 1959.

[43] G. Angloher, M. Bruckmayer, C. Bucci, M. Buehler, S. Cooper, C. Cozzini, P. DiStefano, F. Von Feilitzsch, T. Frank, D. Hauff, et al. Limits on WIMP dark matter using sapphire cryogenic detectors. *Astroparticle Physics*, 18(1):43–55, 2002.

[44] R.F. Voss and J. Clarke. 1/f noise from systems in thermal equilibrium. *Physical Review Letters*, 36(1):42–45, 1976.

[45] J. Pelz and J. Clarke. Dependence of 1/f noise on defects induced in copper films by electron irradiation. *Physical review letters*, 55(7):738–741, 1985.

[46] DM Fleetwood and N. Giordano. Direct link between 1/f noise and defects in metal films. *Physical Review B*, 31(2):1157–1160, 1985.

[47] D. Fleetwood and N. Giordano. Effect of strain on the 1/f noise of metal films. *Phys. Rev. B*, 28(6):3625–3627, Sep 1983.

[48] DE McCumber. Effect of ac Impedance on dc Voltage-Current Characteristics of Superconductor Weak-Link Junctions. *Journal of Applied Physics*, 39:3113, 1968.

[49] WC Stewart. Current-Voltage Characteristics of Josephson Junctions. *Applied physics letters*, 12:277, 1968.

[50] J. Clarke. Principles and applications of SQUIDs. *Proceedings of the IEEE*, 77(8):1208–1223, 1989.

[51] F. Pobell. *Matter and methods at low temperatures*. Springer Verlag, 2007.

[52] WA Little. The transport of heat between dissimilar solids at low temperatures. *Canadian Journal of Physics*, 37(3):334–349, 1959.

[53] FC Wellstood, C. Urbina, and J. Clarke. Hot-electron effects in metals. *Physical Review B*, 49(9):5942–5955, 1994.

[54] M. Kurakado. Possibility of high resolution detectors using superconducting tunnel junctions. *Nucl. Instr. and Meth*, 196:275–277, 1982.

[55] SB Kaplan, CC Chi, DN Langenberg, DJ Scalapino, JJ Chang, and S. Jafarey. Quasiparticle and phonon lifetimes in superconductors. *Physical Review B*, 14(11):4854–4873, 1976.

[56] PCF Di Stefano, T. Frank, G. Angloher, M. Bruckmayer, C. Cozzini, D. Hauff, F. Proebst, S. Rutzinger, W. Seidel, L. Stodolsky, et al. Textured silicon calorimetric light detector. *Journal of Applied Physics*, 94:6887, 2003.

[57] W. Westphal. Development and Characterization of Cryogenic Detectors for the CRESST Experiment. *PhD Thesis, Technische-Universität-München*, 2008.

[58] J. Schmaler. Analysis done at MPI. 2009.

[59] F. Petricca and K. Schäffner. Measurement done at LNGS. 2009.

[60] S. Koynov, M.S. Brandt, and M. Stutzmann. Black nonreflecting silicon surfaces for solar cells. *Applied Physics Letters*, 88:203107, 2006.

[61] S. Koynov. Measurement done at WSI. 2009.

[62] I. Bavykina. Investigation of $ZnWO_4$ and $CaMoO_4$ as Target Materials for the CRESST-II Dark Matter Search. *PhD Thesis, Ludwig-Maximilians-Universität München*, 2009.

[63] B. Cabrera, RM Clarke, P. Colling, AJ Miller, S. Nam, and RW Romani. Detection of single infrared, optical, and ultraviolet photons using superconducting transition edge sensors. *Applied Physics Letters*, 73:735, 1998.

[64] C. Kittel. Introduction to Solid State Physics, 1976.

[65] N.W. Ashcroft and N.D. Mermin. Solid State Physics (Saunders College, Philadelphia, 1976). *International edn.*

[66] W. L. McMillan. Transition temperature of strong-coupled superconductors. *Phys. Rev.*, 167(2):331–344, Mar 1968.

[67] C. Enss and S. Hunklinger. *Tieftemperaturphysik*. Springer, 2000.

[68] W. Buckel and R. Kleiner. *Supraleitung: Grundlagen und Anwendungen*. Physik Verlag, 1977.

[69] BB Triplett, NE Phillips, TL Thorp, DA Shirley, and WD Brewer. Critical field for superconductivity and low-temperature normal-state heat capacity of tungsten. *Journal of Low Temperature Physics*, 12(5):499–518, 1973.

[70] D. Hauff. Experiment done at MPI. 2009.

Acknowledgments

First, I would like to thank Prof. Dr. Allen Caldwell for giving me the possibility to realize this work. Together with Dr. Franz Pröbst, he provides a working place which allows to develop and follow own ideas without barriers. And this in times in which universities adapt more and more to schools, affecting even a PhD student. Being protected from this has been a great benefit for myself. Thank you very much for this ideal scientific liberty!

Beside this, I have to thank Dr. Franz Pröbst for many, many other thinks, especially for the very fruitful discussions and for giving always a second point of view onto a problem. He is an oasis in the world of science. Without him nothing, with him everything is possible.

Dr. Wolfgang Seidel I would like to thank for his always realistic point of view. Unfortunately, this often ends up in a pessimistic view of reality, longing for a more ideal and optimistic world.

Next, I would like to thank Dr. Federica Petricca. She is a (sometimes too) close to perfect person ;) She is very accessible to new ideas and point of views, even when contradicting her own opinion. Without any personal problem she is able to readjust her world view instead of trying to readjust the world to her view.

I do not want to pass over the second Spanish lady (Federica is the first!), Dr. Emilija Pantic. Her work has been the fundament of my work. Beside this, I will never forget her emotional and powerful way of life. Just right to blow with fresh air into the sometimes too dusty corners of science. Stay as you are!

For his diplomatic and the fruitful discussions I would like to thank my PhD student Jens Schmaler. Together, we have been able to solve many scientific and technical problems (which led probably to even more problems). Anyhow, a problem shared is a problem halved!

Thank you Dieter Hauff, for providing me many of the labor intensive basics, like thermometer films or cryostat support, necessary for being able to do research in this field.

Many thanks to Karoline Schäffner! Busy as a bee, she works exemplary and selfless for the CRESST-II experiment. Although you do not think so, this counts very much! Always be optimistic!

Thanks to Quirin Kronseder for the quenching factor basis of chapter 3. Do not doubt! You have the right accuracy for scientific work.

Thank you Michael Kiefer, especially for your permanent computer support. Beside this you are an enrichment for your group.

Also many thanks to Dr. Irina Bavykina, an example for never giving up.

Many thanks to all other CRESSTians, as Dr. Antonio Bento, Rafael Lang, Baojun Luo, and Raphael Kleindienst, for this nice working atmosphere.

Beside the scientific staff of your group, I would like to thank all the members of our collaboration partners, the list of which would blast this section.

Another group I have to thank for the success of this work, are the people from the different departments of our institute. These are beside Hans Seitz people as Peter Mühlbauer, Walter Kosmale, Rainhard Kastner, Gerhard Ott, Hermann Wenninger, Michael Reitmeier, Günther Tratzl, Siegfried Schmidl, Werner Haberer, Trung-Si Tran, Uwe Leupold, and many more people.

Again, many thanks for the proof-readers of this work (If you find a mistake in this work, do not hesitate to contact one of them!). These have been: Jens Schmaler, Dr. Federica Petricca, Dr. Franz Pröbst, Dr. Leo Stodolsky, and Prof. Dr. Allen Caldwell. I hope, these 150 pages have not only been painfull, but also informative for you, so that your many hours have not been only wasted time.

Ganz herzlichen Dank auch an meine Familie, die mich bei all meinen Taten bedingungslos unterstützt!

I want morebooks!

Buy your books fast and straightforward online - at one of the world's fastest growing online book stores! Environmentally sound due to Print-on-Demand technologies.

Buy your books online at
www.get-morebooks.com

Kaufen Sie Ihre Bücher schnell und unkompliziert online – auf einer der am schnellsten wachsenden Buchhandelsplattformen weltweit!
Dank Print-On-Demand umwelt- und ressourcenschonend produziert.

Bücher schneller online kaufen
www.morebooks.de

OmniScriptum Marketing DEU GmbH
Heinrich-Böcking-Str. 6-8
D - 66121 Saarbrücken
Telefax: +49 681 93 81 567-9

info@omniscriptum.com
www.omniscriptum.com

Printed by Books on Demand GmbH, Norderstedt / Germany